# Residential Water Problems
## Prevention and Solutions

Alvin M. Sacks

HOME BUILDER PRESS

Home Builder Press®
National Association of Home Builders
1201 15th Street, NW
Washington, DC 20005-2800

### Dedication

To Charles Martin, deceased; Paul Blake of The Martin Company, Waterproofing Contractors of Washington, D. C.; and Sybe K. Bakker, retired roofing and waterproofing consultant. From them I have learned more about building water problems than I can recount.

**Residential Water Problems: Prevention and Solutions**

ISBN 0-86718-395-0

© 1994 by Home Builder Press®
of the National Association of Home Builders
of the United States of America

Printed in the United States of America.

**Library of Congress Catalog-in-Publication Data**

Sacks, Alvin M., 1927-
    Solutions to residential water problems / Alvin M. Sacks.
      p. cm.
    Includes bibliographical references and index.
    ISBN 0-86718-395-0 : $20.00
    1. Dampness in buildings. 2. Waterproofing. I. Title.
TH9031.S23 1994
693'.892—dc20                             94-15449
                                               CIP

### Disclaimer

Home Builder Press, the National Association of Home Builders, or any other organizations or persons named in this publication offer no warranty nor representation that this information is suitable for any general or particular use nor do they provide freedom from infringement of any patent or patents. Any individuals making use of this information assume all liability from such use.

This publication is designed to provide accurate and authoritative information in regard to the subject matter covered. It is sold with the understanding that the publisher is not engaged in rendering engineering, legal, accounting, or other professional services. If engineering or legal advice or other expert assistance is required, the services of a competent professional person should be sought.
—Adapted from a Declaration of Principles jointly adopted by a Committee of the American Bar Association and a Committee of Publishers and Associations

For further information, please contact—

Home Builder Press®
National Association of Home Builders
1201 15th Street, NW
Washington, DC 20005-2800

*8/94 HBP/Data Reproductions 2.5M*

# Contents

# Figures

# About the Author

Alvin M. Sacks, president of Alvin Sacks, Inc., is a construction consultant and arbitrator with 20 years experience building and remodeling houses and developing land. In 1976 he became a private building inspector, and two of his current specialties are leakage and drainage problems. He lectures to contractors and other groups and has testified numerous times on construction matters.

Sacks was one of the principal authors of *Standard Guidelines for . . . Urban Subsurface Drainage* published by the American Society of Civil Engineers. He has also written dozens of articles for national magazines and newspapers. He is a former officer of Suburban Maryland Building Industry Association and a member of the American Society of Home Inspectors, Roof Consultants Institute, and the American Society of Civil Engineers. He is licensed as a private pilot.

# Acknowledgments

I thank Larry Johnson, Soil Scientist, Fairfax County, Virginia, with whom I have had many conversations over several years about soils and water; Howard R. Sacks, retired Dean and Professor of Law, School of Law, University of Connecticut, West Hartford, Connecticut, my older brother, who encouraged me to write this book and who critiqued the outline; Claxton Walker, retired home builder, remodeler, and building inspector from whom I learned the value of technical writing.

The following persons reviewed both the outline and the book manuscript: James "Andy" Anderson, President, Anderson Homes, Goose Creek, South Carolina; Barbara Cook, PE, Associate, Dames and Moore, Bethesda, Maryland; Bill Eich, President, Bill Eich Construction, Inc., Spirit Lake, Iowa; Susan M. King, PE, King Consulting Engineers, West Seneca, New York; and Chuck Moriarity, President, Moriarity and Matsen, Seattle, Washington.

Reviewers of only the outline included Franchot L. Fenske, PE, Partner/Senior Director of Municipal Engineering Services, R. W. Beck, Seattle, Washington; David E. Beck, PE, Senior Marketing Specialist, Contech Construction Products, Inc., Middletown, Ohio; and Richard Morris, Senior Advisor, NAHB Energy and Home Environment Department.

James V. O'Connor, City Geologist, Department of Biological and Environmental Science, University of the District of Columbia, Washington,

D.C., reviewed the appendix, and R. Gregory Hamadock, Laboratory Manager, Law Engineering, Chantilly, Virginia, reviewed the information on soil mechanics.

# Book Preparation

*Residential Water Problems: Prevention and Solutions* was produced under the general direction of Kent Colton, NAHB Executive Vice President, in association with NAHB staff members Jim Delizia, Staff Vice President, Member and Association Relations; Adrienne Ash, Assistant Staff Vice President, Publishing Services; Rosanne O'Connor, Director of Publications; Doris M. Tennyson, Director, Special Projects/Senior Editor and Project Editor; David Rhodes, Art Director; John Tuttle, Publications Editor; Carolyn Poindexter, Editorial Assistant; Vivian Moore, Word Processor; and Rachael Hymas, Intern.

# Sources of Water and Control Methods: An Overview

## Precipitation

Precipitation in the form of rain, fog, snow, and sleet falls onto roofs, balconies, decks, pavements, and terraces of homes and on the land around them. Rain and snow also blow against vertical surfaces such as frame, masonry, and glass walls. Where the precipitation directly strikes a building, it must be kept outside whatever the surface.

The runoff from precipitation should be controlled to make a building watertight. The devices and methods to make it watertight also require maintenance over the life of the building to keep it watertight.

### Flashings and Other Devices

The joints between surfaces of materials are protected by *flashings* such as *caps and copings*, valleys, and waterproof membranes. Metal cap flashings typically cover the top edges of shed roofs where they extend beyond an adjacent surface. Metal copings typically cover parapet walls. (Sample flashings appear in Figure 1-1. Terms in italic are defined in the glossary.)

Metal flashing and heavy waterproof sheeting provide protection for valleys of roofs. The metal or membrane may be exposed or covered. In cold climates that may subject houses to *ice dams*, a sheeting membrane may cover the roof a few feet up from the eaves under the shingles or other covering. A sheeting membrane is often used under siding to turn out water that penetrates between the boards or panels. At or near *grade* this membrane is especially useful for protecting the framework from water that splashes up from the ground and gets behind the siding. Wind-driven rain also may penetrate lap or board siding.

Water that runs down the face of siding may be turned out by a wood *water table* (Figure 1-1) that is laid on top of the foundation and under the first siding board or panel. While it may slope like a window sill and have a drip edge underneath, water tables frequently rot and permit runoff to get behind (on the inside of) the foundation wall.

To control the water running off the eaves of various roof configurations and slopes (pitches) builders and remodelers use such devices as gutters, *diverters, gravel stops* (metal edge terminations), and *drip edges* (Figure 1-1). These and other devices collect the runoff from surfaces and safely dispose of it via valleys, gutters, and downspouts (leaders) to other surfaces or to the ground. Because gutter purposes and shapes are well-known to readers, gutters will not be described here.

Diverters are flat metal strips a few inches wide and perhaps a foot long. They are often attached to the top edge of a gutter, as a "side board," to direct the water flowing down a valley away from the gutter intersection to prevent spill-over during a hard rain. In this case they are shaped into an *L* with both legs of equal length. A diverter also is used on a gutter that

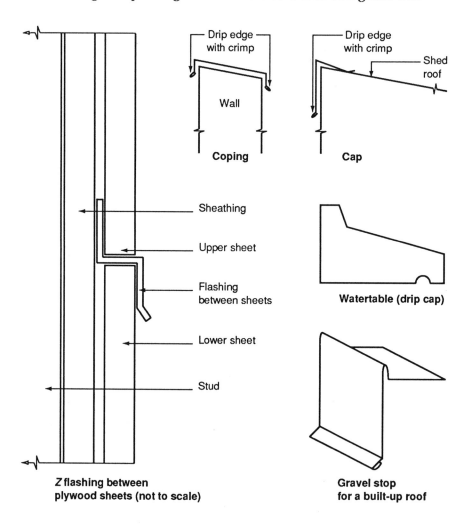

*Figure 1-1. Sample Flashings*

receives concentrated flow from a downspout that dumps higher up on a shed roof.

Gravel stops (Figure 1-1) usually are used at the edges of low-pitched roofs (a) to retain on the roof surface most of the gravel that might be moved toward the edges by the flow of water and (b) to direct the flow of runoff toward a *scupper* or gutter. Scuppers (Figure 1-1) are short openings through flashings or low walls, such as roof parapets and balconies, to permit water to run off the roof or deck before it ponds too deeply and causes a structural hazard because of its weight. Usually roof scuppers are metal channels or pipes that discharge into gutters or downspouts. But at balconies, porches, and the like, scuppers may just be a brick left out of the wall (also known as a weep). Gravel stops come in different heights appropriate to the design of the roof and its drainage.

Drip edges (Figure 1-1) are formed at the bottom edge of the vertical leg of gravel stops, caps and copings, window and door sills, and at many other locations to prevent overflowing water from curling back toward and down vertical surfaces. (The water's strong *capillary tension* causes it to curl back.) Sometimes the drip edge is formed by the metal jutting out at an acute angle; other times it is formed by an upside-down *U* rabbeted into the bottom of a block-shaped surface.

Another common flashing used at intersections of exterior plywood wall panels is in the shape of a double *L*. It is often called a *Z* flashing (Figure 1-1.) At horizontal joints one leg of it tucks behind the edge of the upper piece of siding, and the other leg covers the edge of the lower piece. The leg covering the lower piece should end in a drip edge as discussed above.

The water collected in gutters flows through downspouts of various shapes, sizes, and materials down the sides of buildings. As a substitute for downspouts, heavy chains have been used on the Olympic Peninsula of Washington state. From the gutter the water flows down in and out of the chain's links so that much of the runoff is slowed enough to lose some of its force by the time it spills onto a splashblock (splashpan).

Downspouts end near grade in *shoes* (the curved piece at the bottom of the downspout) that empty into splashblocks at grade. Water should then flow away from the building on the sloping backfill and out onto the yard toward a proper disposal, such as a storm drain or street gutter. Underground piping to a safe outlet may be substituted for downspout shoes and splashblocks, but it is expensive and subject to future blockage and breakage (Figure 1-2). A safe outlet should not include *drywells* unless (a) the drywell is far from the house, and (b) it has a *relief* device discussed in Chapter 4.

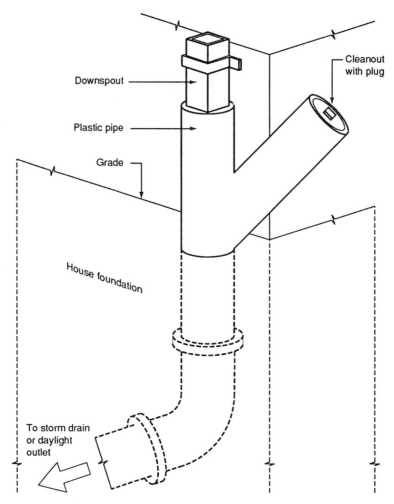

*Figure 1-2. Underground Piping*

*Roof water can be routed to a stormwater drain or other outlet.*

Source: Richard H. Rule, "Making Basements Dry," *Home and Garden Bulletin*, No. 115 (Washington, D.C.: U.S. Department of Agriculture, 1970), p. 4.

## Yard Grading and Landscaping

Runoffs from various surfaces of a home must be disposed of after they land on the ground at or near the foundation. Just sloping the backfill area that will become a grassy or planted surface usually is not enough. The simplest situation has the yard sloping away from the house in all directions. But where the *grading* is more complicated, an overall plan is necessary to combine the different slopes into a long-term, workable solution.

At one extreme a flat yard will have little or no erosion, but it will often pond water for many hours after precipitation ends. At the other extreme a steeply sloped yard quickly delivers runoff to the toes of slopes, but along the way the water often cuts *rills* or *gullies* into the ground (Figure 1-3).

Various grading (land-forming) methods or devices that are more complex than straight slopes are used to divert or direct flows across a yard). A *swale* collects runoff and directs it in a wide, shallow channel downgrade. Its advantage is an unobtrusive appearance that is easy to mow (Figure 1-4).

Sheet erosion

Rill erosion

Gully erosion

Stream flow

*Figure 1-3. Three Phases of Erosion*

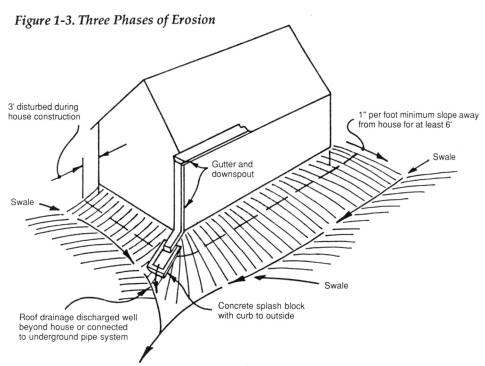

3' disturbed during house construction

1" per foot minimum slope away from house for at least 6'

Swale

Gutter and downspout

Swale

Swale

Concrete splash block with curb to outside

Roof drainage discharged well beyond house or connected to underground pipe system

*Figure 1-4. Idealized Grading Around the House*

*The gutter and downspout in the drawing are joined in a stylized (not real) connection.*

Source: Adapted with permission from Figure 1-3, "Grading Around the House," Stephen J. Spano and Douglas N. Isokait, *Residential Drainage* (Laurel, Md.: Washington Suburban Sanitary Commission, 1979), p. 6.

A *berm* (Figure 1-5) or *dike* intercepts cross-flows and turns them down-grade. Since this ridge-like device would protrude above the grass level, its appearance is undesirable in an open yard. And most mowers will scalp it. However you can use a berm, dike, or *diversion* in a wooded or other grass-free area. Among trees it may follow a curving path to protect their roots, and it is often oriented diagonally down across the contours of a slope to improve the flow characteristics of the runoff.

A diagonal diversion works in two ways. First, it slows the water running off the slope compared to the water in a channel that is directed straight down the slope. Second, because water enters first at the high (upper) end of the channel it creates a momentum to push ahead of it each new volume as the inflows enter farther down the channel at a later time. This hydraulic characteristic allows the diversion to have a reduced slope across the contours. The slope should not be so gentle, however, that rainfall in a heavy downpour will spill out of the channel before it can flow away. Other grading devices include *level spreaders* and *terraces* (Figures 1-6 and 1-7).

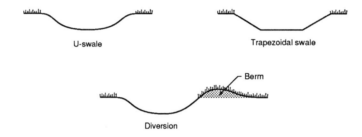

*Figure 1-5. Swales and Diversion*

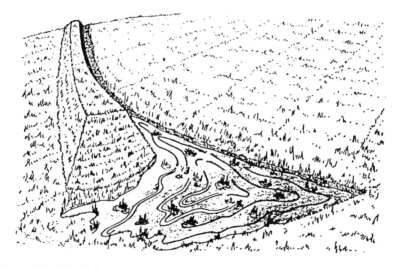

*Figures 1-6. Level Spreader*

Source: Reprinted with permission from *Erosion and Sediment Control Handbook*, 2nd ed. (Richmond, Va.: Soil and Water Conservation Commission, Commonwealth of Virginia, 1980), p. III-161.

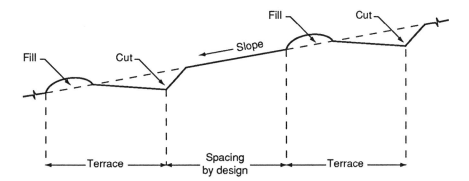

**Figures 1-7. Terraces**

## Erosion Reduction Methods

Erosion reduction—rather than the ideal control—takes place when you significantly reduce the velocity of water because this change substantially decreases its force. Methods to reduce erosion and *sedimentation* divide into structural and vegetative.

Structural methods include the grading devices mentioned above plus various walls such as *revetments, toe walls,* and *retainers* that are built of various materials such as masonry, concrete, stone, and wood. It also includes protection at pipe outlets. A revetment is a mechanical ground cover that (a) armors a slope that is too shallow to require a retainer or (b) forms an apron over flat ground. An example is rocks at the end of a pipe outlet. The rock apron dissipates the energy of the flowing water, slowing it, and allowing some of the silt carried in suspension to settle out (*sedimentation*).

A *stilling basin* adds curbing or low walls on all sides of the apron. These barriers temporarily collect and delay the water for a short time until it overflows at one or more places (*detention*). An opening along one side of the wall or curbing releases the ponding water in a preferred direction, such as into a channel.

When fast-moving water is forced to "climb out" of a pond and over a barrier, a form of hydraulic jump occurs. The accompanying reduction in velocity dissipates some of the energy contained in the water, thus decreasing the water's scouring action beyond the barrier. Probably the best example of this hydraulic jump is a concrete splashblock laid with the curb end farthest from the downspout shoe (Figure 1-8).

Toe walls are a cross between steep revetments and retaining walls. Only 1 or 2 feet tall, they are located at or near the feet of slopes that can be mowed or planted with shrubs.

A retaining wall creates a transition in grade where land forming alone would leave a slope too steep to walk. It must be strong enough to resist the overturning pressures from the soil and water behind it. This risk grows as the wall height increases. Tall walls are expensive and unpleasing to the eye when they are built across a visual path. Low border walls are useful for separating planters from grassy areas, but when they are built solid and

*Figure 1-8. Splashblock Used as a Hydraulic Jump and Combined with a Stilling Basin*

*During a rain the outflow from a 2x3-inch downspout (unseen at right) flows less than half full. After the outflow climbs over the curb at the end of the splashblock (which creates a hydraulic jump), it is stilled by unseen rocks beyond the end of the splashblock. From this stilling basin the outflow turns to flow over and through additional rocks and then spreads outward at the level ground beyond. The brickwork is the top of the adjacent areaway wall.*

close to homes, they often impede the flow of water away from the foundation.

Temporary structural devices for slowing surface flows include plastic *silt fences*, staked plywood *velocity checks (check dams)*, straw bales, and piles of rocks. Individually these devices serve one of two purposes: they attempt to separate out and detain the *fines* (fine particles of soil) carried in suspension by fast-flowing water), or they should reduce the velocities of runoff by making it flow over and through tortuous and rough paths. (For the difference between *detention* and *retention* see the glossary at the end of the book.)

Vegetative methods for reducing erosion and sedimentation include plant materials in hedges or beds, ground covers, and direct implantation of fast-growing live tree twigs and branches that will root quickly. Around finished areas, you would still use shrubs and ground covers to protect slopes.

# Groundwater

Water from a second external source may attack homes. *Groundwater* may penetrate basements and crawl spaces or saturate grassy and wooded areas around homes. (As noted in the glossary, this word has different meanings.) The word is used here to include all subsurface flows from wherever they originate. Different origins include a *water table* condition, water seeping underground from an overflowing dry well, water *recharged* into the ground via roof runoffs, and others.

You can control the attacking groundwater by intercepting it, collecting it, and conveying it to a safe disposal. While these three purposes may be separate in function, usually they are combined into an integrated system. Interception is accomplished by using a ditch that is at least filled with gravel, a *French drain*, and more often also includes perforated pipe, making it a *trench drain* (Figure 1-9). Both these drains also collect and convey water, but for conveyance only, solid pipe is preferred and the gravel *envelope* is minimized.

*Geotextiles*, such as filter fabrics, often have been used in the past to enclose the gravel of an underground collection system to reduce the *clogging* hazard of slowly migrating fines such as clayey soils. For silts and sands that easily liquefy, current practice is to skip the fabric and rely only on the gravel filter. This procedure may change again in the future, so you need to watch for new research results.

The location of underground drainage depends on where the subsurface flows can be intercepted. At least several feet from the house is preferable because the distance will reduce the combined pressure of the water and soil on the foundation wall. Water weighs 62.4 pounds per cubic foot and applies equal pressures in all directions. Soils average less than 100 pounds per cubic foot and apply their pressures primarily in the direction of gravity. Water and soil together exert a lateral pressure component on walls that averages 30 to 40 pounds per square foot. Both the water and soil weights are based on volumes 1 foot tall.

If water saturates the soil, for example, because the drainage system has failed, the total pressure escalates perhaps to three times the dry soil pressure acting alone. This excessive pressure often breaks both foundation and retaining walls.

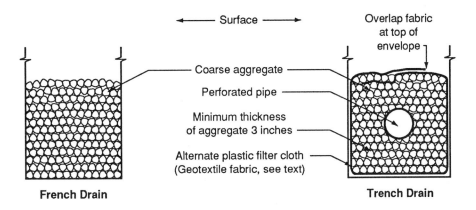

**Figure 1-9. French Drain and Trench Drain**

*Filter cloths may be used with a gravel envelope.*

Source: Adapted from drawing courtesy of the Soil Conservation Service, U.S. Department of Agriculture, Washington, D.C.

## Underdrains

If the groundwater source cannot feasibly be located away from the building, *underdrains* are installed on the outside or inside of foundations. Many new house builders prefer the pipes on the inside because they are better protected from trash in the backfill and are easier to flush in the future.

Underdrains lie along or on top of the footing shoulder where they are surrounded by gravel that is covered, but not wrapped, by a membrane to reduce the infiltration of soil into the gravel. If these systems are placed outside, they should be hand backfilled for the first couple feet of cover.

## Dewatering Devices

When underdrains cannot be drained by gravity to a safe outlet, you may need to pump the water from a collection pit (sump). This pit may be a gravel-filled hole, a flue liner with holes through the walls, or a plastic sump pit with knock-outs for the drainage pipes.

Deciding whether to locate the pit in a basement depends on whether the area is to be finished and on where the low point of the outside grade is. For cosmetic reasons a sump pit should be located in an unfinished area. Water exiting at the high grade corner of the foundation may just flow downhill as it follows the slope of the backfill instead of flowing away from the corner. Or sometimes the runoff will start flowing away from that corner and then curl back toward the foundation where it ultimately enters the wall at some point.

If you pipe the water from the sump pump to the outside, you should direct it away from the foundation onto a splashblock or empty it into a pipe with adequate fall and with the outlet well away from the house. When the outlet pipe exits the building above grade without such protection, the high velocity flow out of a sump pump usually causes erosion. You should put an airtight lid on the sump pump to prevent soil gas, radon gas, and water vapor from getting into the basement.

## Membranes

Additional devices for keeping water out of a building include *impermeable membranes*. These membranes are used on the outside walls of a building on the exterior (*positive side*) or the interior (*negative side*) of the wall. The positive side works better because the water attempting to penetrate is not allowed to enter the wall. If water enters the wall and is trapped by a negative side membrane, the water may ultimately damage the wall. While such damage is relatively rare in some geographic areas, in other areas with chemically aggressive water, walls are susceptible.

Builders and remodelers also use these membranes under the slab of a basement or a crawl space. Take care to prevent punctures during the laying of steel reinforcement and placing the concrete over the membrane.

## In-Plane Drains

In especially deep basements with tall walls in areas in which groundwater is an ever-present hazard, you could use an additional device together with, or as part of, the waterproof membranes attached to the positive side of the walls. Called an *in-plane* drain, it consists of plastic cores embedded in a plastic sheet. The plastic cores are made in different shapes by various manufacturers to intercept free water. The intercepted water passes down the wall and into underdrains. This system prevents the buildup of water pressure that could separate membrane seams and get behind the fabric.

Installing the basement floor on a 4-inch layer of crushed or washed gravel allows water under the slab to flow more freely to the drain tile. Aggregate also forms a capillary moisture break to prevent water from wicking or seeping up through the basement floor.

# Plumbing, Heating, and Cooling Lines

The various pipes that supply and drain domestic water and that heat and cool homes are a third source of water inside homes. Supply pipes from outside sources (such as public utilities and private wells) may pass through meters before joining interior and exterior supply lines that are connected to fixtures inside and outside the home. Because they are under pressure, they are more likely than drains (discussed in the next paragraph) to develop leaks. Such leaks often occur at connections and fittings, but they also may occur in the wall of a pipe.

Drains and waste lines inside and outside homes carry "black" water from toilets, and "gray" water from the other fixtures. These drains and waste lines should be designed and constructed to prevent the buildup of excess pressure within individual pipes that could (a) cause leakage at weak connections or (b) split pipes open. A venting system connected to drains and wastes helps to relieve pressures and prevents internal siphonage from traps.

Heating pipes carry hot water or steam from boilers to radiators and other terminal devices. The returned water or condensate flows through pipes back to the boiler, sometimes in a complicated manner. Some systems are quite sophisticated and include accessories that require periodic maintenance if they are not to leak or unnecessarily shut-down the system.

Cooling pipes may carry plain water or brine from refrigeration equipment to *air handlers* and back.

Many pipes in homes are concealed in walls, ceilings, and floors. Sometimes they are buried inside or under concrete slabs. These pipes are subject to the leakage discussed above, plus the problems caused by corrosion of certain metals embedded in concrete.

# Condensation

A fourth source of water inside homes results from the physical process that occurs when vapor cools below its *dew point* and incrementally releases some of its invisible water in the form of suspended droplets. The cooling takes place in two major ways: from the loss of heat to the outside during wintertime and from the transfer of heat to the outside by mechanical cooling of interiors during summertime.

Symptoms of condensation are obvious when moisture appears on windows, bathroom surfaces, drinking glasses containing iced drinks, and refrigerator surfaces near the edges of doors. However, symptoms of condensation may be invisible and perhaps more serious on insulation inside walls, above ceilings, and below floors.

You can control condensation by using vapor and liquid retarders and barriers such as sheet plastic and rubber-type membranes, asphalt-impregnated sheathing, and other building wraps. You can also solve condensation problems by (a) reducing excess humidity with kitchen and bathroom vent fans or whole-house, central ventilating systems and (b) limiting the output of the source of excess humidity, for example, by resetting an overactive humidifier, drying out a wet basement, covering an unpaved crawlspace with a vapor retarder, removing excess plants, and venting the clothes drier.

# Prevention of Foundation Leakage Before Construction

Developers, builders, and contractors should study sites for potential water problems before they purchase land, or at least do so before starting to excavate each lot. For water problems in houses (especially in their basements and crawl spaces), the cost of correcting a problem that could have been prevented could be three or more times the cost of doing it right the first time because you would have to add the cost of later demolition and replacement to the original cost of installation.

For developers buying raw land, builders buying finished lots, and remodelers adding to existing homes, the following paragraphs provide procedures for studying a site for potential water problems. These procedures include researching documents related to the site, on-site investigation of the surface and subsurface of the site (examining the characteristics of the plants, soil, and rock at the site), and treatment of potential water problems before construction.

## Documents

A paper search usually is the first step toward discovering potential or actual water problems prior to studying the actual site. The documents listed below might be among those available from many public and private sources:

- topographic maps, boundary surveys, old photographs, and land records
- municipal building permits, engineering, and transportation departments
- state and local studies of geology, hydrology, drainage, and soils
- county soil surveys conducted by the Soil Conservation Service, U.S. Department of Agriculture
- old records and charts of mines, water sources, abandoned foundations, garbage dumps, and the like
- street plats and utility company drawings
- newspaper morgues (files of old articles and photographs)

- old photographs from local historical society, library, photographers, and history buffs
- old postcards showing the site

Unless you are already skilled at interpreting the various documents listed above, you probably will need help (a) learning to read symbols and layouts and (b) understanding the similarities and differences between and among materials and formations. Specialists who should be able to help you include geologists, hydrologists, agronomists, soil scientists, a broad range of civil engineers, erosion control and drainage consultants, real estate agents who specialize in land sales, surveyors, well drillers, experienced builders and general contractors, neighboring landowners, librarians, local officials, and others.

# Soil and Rock Characteristics

If not anticipated, underground conditions, such as a high *water table* or hard rock, could ruin a building budget, price a speculative house way above the market, or create a big problem for a customer whose loan is at the ceiling of the customer's ability to repay it. Knowing about soil and rock characteristics enables you to make judgments about the seriousness of a particular situation and to determine whether you should go ahead with the job or bail out early. Although you cannot ignore the legal implications of quitting, wasting money by continuing may be an unwise alternative.

Most of us would agree that soil and rock are different, but occasionally soil looks and acts like rock, and vice versa. For example, a soil *fragipan* (an extremely firm and confining layer of variable depths) may be so hard that water will not penetrate it. One type of rock with some of the characteristics of soil is sometimes called rotten rock because it is soft and fragmented. The selected bibliography at the end of this book lists some texts on rock and soils for additional reading.

Professionals prefer the word *soil* to *dirt* or *earth* because it's more scientific and all inclusive. For example, the term *soil excavation* also includes excavation of rock. *Dirt* is a lay person's term.

Modern soil management, an application of soil science, concerns itself with different characteristics of various soils. While soils have many characteristics, the important ones for management include fertility and *tilth*, drainage (internal and external or runoff), profile (horizons), topography (land forms) and *relief*, color, *texture, structure, consistency, reaction*, erosion, and classification.

Thousands of soils have been identified and classified. Knowledge of these characteristics and classifications can help builders and remodelers to guard against water problems by taking early preventive actions both prior to and during construction.

Construction also involves two other subjects—soil mechanics and soils engineering—that are based on studies of soils and geology. Soil mechanics combines the knowledge gained from solid and fluid mechanics with studies of strength of materials and applies them to soils. Soils engineering applies soil mechanics to practical engineering design and construction problems, such as foundation sizes and depths, retaining wall materials and sizes, slope stability, and structural settlement.

## *Soil's Attraction for Water*

For the purposes of this book, soil's affinity for water is one of its most crucial characteristics. Does water pass through the soil quickly, or is water held in the soil's pores (*absorption*) and/or on the surfaces of its grains (*adsorption*)? Three words help to describe this affinity, *porosity, permeability,* and *capillarity.*

*Porosity* refers to the air (or void) spaces between solid particles. As a percentage, it is defined as the volume of air spaces compared to the total volume of the sample.

*Permeability* (sometimes called *hydraulic conductivity*) refers to a material's ability to allow the free flow of water and air through its pores in any direction. Permeability is affected by a material's porosity, grain size, and structure.

*Capillarity* is the rise of water through tiny pores in a soil or other materials such as tree fibers. Capillarity is caused by molecular attraction forces such as seen in *surface tension.* (In solid materials the words *adhesion* and *cohesion* may be used.) Capillarity also occurs in man-made materials such as brick and block.

## *Natural Soils*

Examples of natural materials in the ground and how they pass or hold water, include clay, silt, organics, sand, loam, gravel, and rock. The preceding names are sometimes used to describe the sizes and the composition of various materials. In the following discussion the names will apply to both size and composition. Several authorities promote size classification systems. This book will use the American Society for Testing and Materials standard for engineering purposes—D 2487-92.)

*Clays* are chemically complex minerals and the finest particles of soil presently known. They occur in different colors and are easily identified by their stickiness when wet and their hardness when dry. Because their particles are microscopic—less than 0.003 inch in diameter—they are packed extremely close together. Their porosity may be 50 percent (meaning that 50 percent of the volume of a sample would be air spaces). However their air spaces are so tiny that they strongly resist the free passage of water through them. Clay particles also have such a strong chemical attraction for water that they severely retard the flow of water. Nevertheless, capillary movement of water in clays has reached vertical heights of 30 feet or more.

You can use a simple field test to identify clay (Figure 2-1). Pick up less than a handful of the material that is damp enough to roll into a ball without sticking to your fingers. Roll it out between your hands into a thinner and thinner thread. If the diameter of the thread decreases to 1/8 inch without falling apart, the sample has a significant amount of clay in it.

In many parts of the country clays containing the minerals kaolinite and illite may retain water, but usually they do not cause foundations problems from swelling and shrinking. Elsewhere the clays containing the minerals smectite and montmorillonite expand when they are wet and contract

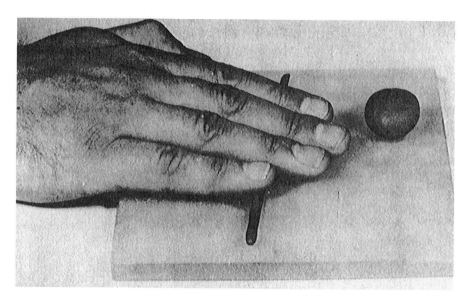

*Figure 2-1. Clay Thread Test Variation*

*The roll or thread test is used in the field to identify soils by determining plasticity. The samples show the rolling process and the ball of soil before it is rolled out.*

Source: *Soils Engineering*, Sec. I, Vol. I (Fort Belvoir, Va.: U.S. Army Engineer School, 1971), p. 54.

when they are dry. These clays have the potential for doing so much harm to a home over the years—through differential heave and settlement—that special foundation methods should be followed during construction.[1]

*Silts* are closest in size to clays and are also less than 0.003 inch in diameter. They differ from clay in their origins and in their low *plasticity*. Pure silts are residues from weathered rock. They look and feel floury and are often called rock flour. Mixed varieties also may contain organic material, clay, sand, and other residues, but silts do not hold water like clays. When dry, silts break easily from finger pressure. Another characteristic of silt, that differs from clay, is its tendency to separate from other particles and to be carried in suspension by flowing water (*liquefy*). Some silts will support capillarity.

*Sand* is a loose granular material that develops from the *disintegration* and *decomposition* of rocks. Sand grain diameters range from about 0.003 inch to less than 0.2 inch. Clean sand has large air spaces (porosity), usually has high permeability, and does not support capillarity.

Organic materials are the products of the decay of plants, insects, and animal life. Examples are top soil, peat, and the humus found under trees. For retaining water, organics generally rank between clays and silts.

*Loam* is a soil mixture containing clay, silt, sand, and organics in differing proportions at different locations. Like organic materials, loam's capillarity and permeability vary, but they generally fall between those of clay and silt.

A triangular chart of *textural* classes (Figure 2-2) distinguishes among the fine sand, silt, and clay contents of a soil sample.

Gravel is larger than sand; its diameters range from 0.2 inch to almost 3 inches. Because of gravel's large void spaces and high permeability, drainage envelopes usually include it.

As part of the earth's crust, *rock* is the original parent of most soils. Rock occurs in large masses, such as bedrock or ledge (a term used in New England), or as fragments. Loose rocks range in size from cobbles (3 to 12 inches in diameter) to boulders (more than 12 inches in diameter).

The ease of rock excavation depends on its structural condition. Soft or loose materials may be dug using a standard front-end loader, tighter formations will require ripping by machines with rock teeth, and the solid faces will require blasting. Rock may be quite hospitable to water, or it may be impermeable. Rock also may contain water that was trapped (*connate water*) when the rock was formed or deposited.

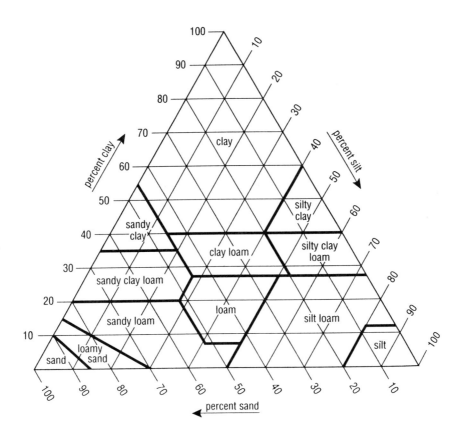

*Figure 2-2. Soil Textural Classes Triangle*

*This triangular chart shows the percentages of clay, silt, and sand in the basic textural classes.*

Source: Soil Survey Division Staff, *Soil Survey Manual* (Washington, D.C.: U.S. Department of Agriculture, 1993), p. 138.

### Problem Soils and Conditions

The list in Figure 2-3 includes some of the local names and descriptions of problem soils and conditions that builders and remodelers may encounter in various locations in North America.

Except for materials such as those listed in Figure 2-3, natural, undisturbed soils usually are of adequate bearing capacity for most houses, but you should watch for those instances when they are not.

# Surface Investigation of the Site

While the paper search (described on the first page of this chapter) is a necessary preliminary, you should walk the land before buying or excavating it. Even then you will not know fully what is under the surface until you complete the excavation, but test pits and/or soil borings reveal useful data as discussed below under Subsurface Investigation of the Site. You have already learned some of what to look for from other people and from documents, but you should keep an open mind and an alert eye for other potential problems. You need to learn enough to react to symptoms on your own and resist being prejudiced by others' opinions, judgments, and conclusions.

As you walk the land, carry and use a checklist (your own or *Land Buying Checklist*.[2] The checklist should contain such items as topography (also called landform or relief), water courses or dry beds, kinds and conditions of trees and plants, wetlands, rock outcroppings, holes, depressions, and odors.

### Topography

While *topography* may be viewed by a site planner as easy or hard, depending on such factors as placement of streets, number of lots, types of houses, trees to be saved, for the purposes of this chapter, you need to look for the location or absence of natural surface runoffs during rain or snowmelt. (For solving water problems, this view is of external drainage not internal drainage or permeability.) Water moves faster on land with high relief than on flat land, but if buildings block the water's paths, the runoffs often penetrate the buildings. On the other hand flat or concave relief allows the water to run slowly or even to pond. If the water accumulates in the wrong places after homes are built, basements and crawl spaces will be threatened with inundation. Three ways of measuring slope are ratio, percentage, and degree of arc (Figure 2-4).

Where slopes are steep, *slope instability* may be a problem merely because of the pull of gravity. The potential for landslides increases when surface runoff and subsurface flows are added to the soil strength equation. If construction occurs the potential is further increased.

If you think that the effect of water on soil is to lubricate it, you would be oversimplifying. Water affects soil strength in a couple of ways: (a) it adds pressure to the mix as the water replaces the air in the spaces between the soil particles, and (b) it introduces chemical matter to the points of contact on the soil particles.

*Figure 2-3. Problem Soils and Conditions Associated with Residential Water Problems*

**bentonite**—A highly expansive clay frequently found in the Rocky Mountains. It is used for waterproofing heavy construction and for drilling mud, but houses should not be built on it without taking the special precautions as cited above under clay.

**bull's liver**—A "quicksilt" that can liquefy (flow) when saturated and vibrated. (See definitions for *quick* later in this list.)

**fuller's earth**—A highly plastic clay that is good for absorbing oils and dyes, but has the same characteristics under homes as bentonite.

**gumbo**—A sticky mud, often of plastic clay, that is found in the Southwest, and along the Missouri and Mississippi River valleys. It is not a good backfill material.

**loess**—Mostly silt deposited by glacial winds and found in the upper Midwest. One example is the bluffs along the Missouri River between Sioux City and Omaha. A peculiar characteristic of this material in a cut is its tendency to slump when rained on, but then to reerect itself into an almost vertical slope when the moisture evaporates. Under and around houses you need to keep loess from ponding water that could soften it and lead to settlement.

**marine clay**—Highly plastic, laid down in saltwater. It is found, for example, in the Interstate 95 corridor between New Jersey and Virginia. As with other extremely malleable clays, building houses on marine clay without using special foundation techniques is risky. If the clay is not too deep perhaps a foundation can penetrate through it to solid material. Experiments with lime as a soil additive have reduced its sensitivity. (Marine clay not formed in saltwater is misnamed.)

**muck**—A general term for extremely soft material that is wet like mud and often contains organics. Obviously it does not belong under footings, so either replace it with soil of adequate bearing capacity or penetrate it with a foundation that reaches such soil.

**peat**—Partially decomposed and putrefied organic matter found in bogs in Canada, Alaska, and elsewhere. Methane gas could be hazardous. Even dewatered, peat's bearing capacity would be less than minimal.

**permafrost**—A condition that occurs in extremely cold climates in which the ground stays frozen all year for a number of years. In the winter permafrost extends from the surface down many feet (the distance depends on climatic and soil variables). In the summer the ground thaws from the surface down variable distances. A building on frozen ground usually settles as the ground under it thaws during the summer, and if the settlement is not uniform, the building suffers structural damage. Over a period of years heat inside the building can radiate downward and increase the depth of the thawed ground and thus worsen the problem. Preventive measures taken during construction include insulation techniques developed at the National Research Center of the National Association of Home Builders in its Frost-Protected Shallow Foundations (FPSF) Program.

**quick**—Before words such as *clay, silt,* and *sand,* it is a condition, not a soil. It refers to the effect of water from any source on the stability of a material. Seepage or hydrostatic pressure causes a susceptible soil to become a liquid. Quick soils are readily erodible. Other symptoms of quick soils include heaves and blisters in the soil as well as boils (rising air bubbles) also called sand boils.

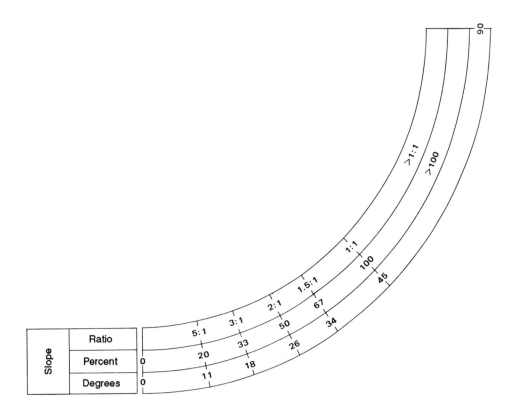

*Figure 2-4. Slope Parameters*

Source: Russell H. Campbell, *Soil Slips, Debris Flows, and Rainstorms in the Santa Monica Mountains and Vicinity, Southern California,* Professional Paper 851 (Washington, D.C.: Geological Survey, U.S. Department of the Interior, 1975), p. 12.

As a practical matter, when rainfall saturates the soil on a steep slope it adds weight to the mass, decreases friction within the mass, and raises the internal pressure—all of which tend to make the slope unstable and subject to failure. Slope movements are relatively easy to see.[3]

## Wet and Dry Water Courses

Water courses show continuing flow while dry beds show where water has flowed in the past and where it may do so again in the future. The crucial characteristics for both water courses and dry beds are their depth, width (not only within the channel but including the flood plain), and their tributaries. Be conservative when you predict how far away to keep each home from major surface flows. To get full use of the land, some minor flows may need to be redirected around or diverted from buildings in their paths.

For large streams you need to look into flood plain restrictions that have been—or probably will be—imposed by public officials. Your choice of building locations may be limited. (This subject will be explored in more detail below in the discussion of subsurface investigation.)

## Water-Loving Trees and Plants

Trees and plants, especially those that draw moisture from the water table or the area just above it. These *phreatophytes* (as they are technically called) provide clues to where underground water can be expected. They include alder, birch, cottonwood, palm, sycamore, willow, ferns, rushes, cattails, and various desert plants. If water-loving trees and plants are not in abundance, you may not react to their presence as you would to a wetland where many of the trees and plants mentioned above grow profusely. Current federal and state regulations make getting permits to work in wetlands quite difficult. Dealing with this problem is a separate subject that will not be covered here.

## Filled Ground

The condition of trees can tell a tale of filled ground. Dead but still-standing trees and those with trunks that do not flair out at grade should raise questions. But if the site has no trees, fill that has only been in place a few years may have an artificial look about it. No matter how long it has been in place and consolidating, you should have fill tested before building on it. Initially this condition is more of a problem for foundation bearing than for subsurface water. However, if the fill is deep, your excavation and subsequent foundation might penetrate the water table and require a dewatering system for the site (a situation that may also occur when no fill is present). Of course, if the fill is so deep that only piles (augered or driven) or drilled shafts are practical and economical, the subsurface water that flows through and around them probably will not be a problem if the first level of the house will be far above the water table.

## Other Clues

Rock outcroppings probably will not point to underground water sources, but they might alert you to the possibility or probability that only shallow foundations will be feasible. If so, subsurface water should be less of a problem than with full basements. If you build in the upper Midwest, Alaska, New England, and elsewhere where glaciers invaded the terrain, you will not mistake *drumlins*, (oval hills of sand and gravel) for rock outcroppings.

Holes and depressions may have water standing in them, or they may indicate where water previously ponded. The water came from surface run-off and/or from subsurface penetrations. You should find out from which source the water came because surface runoffs are much easier to deal with as discussed above under water courses. Subsurface flows will be discussed below.

Holes and depressions in the ground surface that show a vertical displacement along their edges where they meet adjacent soil may be indicators of *subsidence* (settlement). Sinkholes may occur over old mine roofs and shafts, limestone caverns, garbage dumps, landfills, soft soils, and the like. In addition to the added foundation costs for reaching solid bearing, major problems may occur from groundwater and potentially explosive methane gas.

Odors encountered while walking the land (with the exception of skunk odor) often signal the presence of rotting organic matter. Underground odors commonly occur when excavating a foundation or digging a footing near a watercourse where *muck* (mud with highly organic content) is normal. Or they might occur under fill that was spread over existing green growth. But if the odor is close enough to the surface to smell before digging, it is warning you to look for the source. Organic materials have a high potential for continual and extensive settlement.

# Subsurface Investigation of the Site

The immediately preceding paragraphs alerted you to many signs of present or future problems. If you see or smell any of those signs, you should find out more about the underground conditions, at least down to the crawl space or basement depth. You should check the underground conditions in the area where the buildings will sit and under the surface of the land adjacent to them. Site conditions will determine the depth and the distance to explore, as discussed below. The paragraphs that follow describe some methods for exploring underground conditions in soil (but not hard rock). They are ranked by difficulty and expense.

## Test Holes

The simplest is to dig test holes with a post-hole digger. The depth is limited to the length of the shovel arms, so this method is more appropriate for slab houses. As you remove samples lay them on a piece of poly or long plywood in the order in which you remove them. As you dig deeper study the samples looking for problems, such as excessive wetness or standing water, dull-colored (gleyed) soil that might indicate a shortage of oxygen, and soft material that collapses in your hand or that bends without breaking. Compare samples to see if problematic conditions run-out or worsen with depth.

If the problem, such as wet soil or visible water, is within the depth of the proposed foundation, or even a couple feet below it, you can explore deeper using a different method. If you are daring, you could wait until you dig for the foundation to see what you encounter and then decide on the curative procedure. The advantage if you are going to excavate anyway is that the soils exploration can start from the bottom rather than the top of the hole. But this procedure has a potential, major disadvantage: if you hit water in soft soil while excavating with a large machine that starts sinking into a morass, you will need to replace the loader or dozer with a small tracked machine that is lighter and more buoyant. If the substitute still sinks, a backhoe becomes necessary, and if that gets mired you are faced with using a dragline (an earth-moving machine that brings soil from a distance in buckets on a cable that runs over a boom). At that time you should strongly consider abandoning the excavation for a house.

## Augers

Soils engineers and other professionals use an open-flight (helical) auger or an Iwan auger for digging and retrieving soil samples by hand (Figures 2-5 and 2-6). The helical auger is quicker to use because as the tool advances the drilled material slowly and intermittently backs up the hole. It often

*Figure 2-5. An Open-Flight (Helical) Auger*

Source: Soil Survey Division Staff, *Soil Survey Manual*, USDA Handbook No. 18 (Washington, D.C.: U.S. Department of Agriculture, 1993), p. 104.

*Figure 2-6. An Iwan Auger*

*The long blades of an Iwan auger are curved enough to hold a soil sample intact.*

damages the materials it enters, and it cannot be used once it hits water. A strong man can drive the auger and its extensions a dozen feet or more.

The Iwan auger has a partially closed chamber that contains samples as the auger advances. However, because it does not hold much, the Iwan has to be removed from the hole frequently to extricate the sample. Drilling through water should not affect the quality of *cohesive* soil samples that have a binder in them, such as clay. However samples of sand and other noncohesives probably will fall apart as they are brought up.

## Pick and Shovel or Backhoe

Another alternative is to use a pick and shovel or even a backhoe to dig a test hole or pit. Depending on the soil and current OSHA regulations, you probably will be limited to about 5 feet unless you protect against cave-ins. If it is safe to enter, a pit provides the advantage of extra space for entering to observe the sides and for digging deeper at the bottom with a post-hole digger or auger. Observation of the sides provides more information on the contents of layers and the presence of water seeps.

## *Steel Bar*

Another device for probing deeper into the bottom of the hole or pit is a steel bar. Plumbers and utility contractors typically use a half-inch diameter, smooth steel bar pointed at one end to locate old ditches or underground pipes. Limiting it to about 5 feet long helps to prevent getting it stuck. You might sometimes drive it with a hammer for deeper probing. But when you use it to check the strength of soil, you drive it with only one hand. If you can only drive it in a few inches, the soil should be strong enough to carry an average house.

## *Boring or Drilling Equipment*

The previous methods are generally limited to relatively shallow soils and water studies. For deeper explorations use truck-mounted boring or drilling equipment that resembles a well-drilling rig. Interpreting the data may require a soils engineer. If one is not available, a structural engineer might suffice because house loads are relatively light and many structural engineers have some soils engineering training or experience.

Generally most soils in the United States get stronger the deeper you dig. So for most homes in known soils, you usually would not test deeper than footing subgrade unless you encountered a problem, such as soft soil, odorous material that is soft and mucky, organic material, a stream bed, expansive soils, previous construction that had been covered over. However before you dig, if you suspect or see evidence of an underlying soil-bearing problem, such as a garbage dump, landfill, worked-out gravel pit, or water course, you probably would need to explore, but not fully excavate, a few feet below the proposed footing subgrade to confirm minimum *allowable bearing capacity* for an average home (about 1 ton per square foot).

# Modifications Before Construction

Rough-grading the site early takes care of surface runoffs that flow toward your proposed foundation. You should redirect those flows away from the proposed building for two reasons: The obvious one is to keep water out of the hole if you are building a basement. (The basement hole obviously will include footing trenches, plumbing ditches, and the like.) The second reason is that if water still may penetrate the hole from the sides, bottom, or both. Although you probably have eliminated surface runoff as the most likely cause, you still must look for a subsurface cause.

## *Water Tables*

If you see water standing in a test hole before you start to excavate, it may show you how high the water table[4] stands, but which table of the three:

- A relatively shallow *perched table* in your immediate vicinity only occurs after heavy precipitation, or stands seasonally. A perched table stands above the next table over an impervious layer between them.
- The *main water table* extends far and wide in your area and that might be much deeper than a perched table. This one is also known as a *local*, an apparent, or a normal water table.

- The top of an *artesian system* (water under pressure) that could extend much deeper than you will be prepared to dig through or alongside.

## Dealing with Water Table Problems

Because a perched table may not be deep, you might be able to stay above it for a slab house, or in it for a house with a crawl space or a basement if the soil is strong enough to carry the load it ultimately will bear. If the soils are not strong enough to carry the ultimate load of an occupied home, you may have to dig through the wet material to find adequate bearing or choose an alternative foundation system such as piers, piles, or compacted fill.

Obviously, building a house in water will be extremely difficult so you will need to start by dewatering the subgrade. To handle the subsurface water outside crawl-space and basement jobs, you may need to dig a sump pit outside the proposed foundation to lower the water table below the building's subgrade. If the soil is of low permeability so that groundwater flows too slowly to the pit, lay perforated pipe enclosed by a gravel envelope to intercept the water outside the future foundation walls. The gravel should be approximately 2 inches in diameter and without fines. (This structure is intended as a site dewatering system; it is not a house dewatering system. That system will come later during construction, see Chapter 3.)

If the ground has substantial slope down toward the foundation perhaps a straight line pipe laid upgrade from the building will suffice. But if the ground is relatively level around the foundation, the underdrain previously described should circle the future building several feet away from it. You should maintain a "respectful" distance between the underdrain and the footing so that future vertical changes in the water table do not unduly weaken or add pressure to the soil under the footing. Such effects on soil carrying a house load could contribute to future settlement and heave of the building. When the perched table rises after heavy rains or melting snow, the interceptor should dispose of it if the sump pit is automatically pumped out or if one or both ends of the pipe flow by gravity to an adequate outlet(s).

In an artesian situation where you are already at subgrade, you need to move the proposed foundation away from a vertical water column or exposed seep if you can. If you cannot find a better location for the foundation, and you can find a place that has only one or two eruptions of water, "capping" them is an alternative. Capping involves installing a gravel envelope (sides and top) around the issuing water. Drain the envelope by pipe to an adequate outfall outside the foundation or include it in the inside dewatering system described in Chapter 3.

You should not excavate, as opposed to isolated digging, through wet soil until it has been tested to find out how deep you will have to go to reach adequate bearing. Deep boring might not be necessary because the deeper you need to go the more likely a deep foundation, such as piles or poles, will prove to be more economical than regular block, concrete, or wood walls. And their feet still will be under water. Another alternative is a reinforced concrete slab (sometimes called a *mud-slab* or *mat*). Such a slab would have to be strong enough to carry the weight of the whole house and large enough to distribute the whole house load over an area that will have enough bearing capacity at the slab depth.

If you find a groundwater condition but still must go deeper to reach subgrade, you should stop to consider redesigning the project or the foundation. If you can substitute a slab or crawl space for a basement and stay above the water and soft soil, you may not need an expensive substitute foundation and an extensive dewatering system. However, if you want to keep the basement, maybe you can raise the whole house enough for the basement to stay dry. Usually raising the whole house will require raising outside grades, too.

# 3

# Treatment of Foundation Leakage During Construction

Occasionally after a basement or crawl space is excavated and the hole is left overnight (until the footings are started the next day), water will enter the excavation from two natural sources: precipitation or groundwater.

## Precipitation

Once the footings are in place the influx of water may still be a problem, but less of one. Rain or melted snow, sleet, or ice may flow by gravity into the hole. Usually this situation requires only a simple solution such as mounding up the soil near the hole so that surface flows are turned safely away. Additionally, a *swale* or *berm* may be needed farther from the excavation to divert the surface runoff.

## Groundwater

A second source could be groundwater (also known as subsurface or subterranean water as discussed in Chapter 1 and the Appendix). Groundwater may penetrate the excavation even though no rain, sleet, or snow fell. The water is usually clear and may stand only in one corner. It may (a) seep out of saturated soil nearby, (b) issue from a *water table* that was disturbed, or (c) rise in the soil from underground pressure, either natural (*artesian*) or man-made (plumbing). Artesian water flows between confining layers underground and is under heavier pressure than atmospheric. When a confining layer is penetrated water flows out of the hole (also see the Appendix and *water table* in the glossary).

### Seepage

Sometimes you can discover the seepage source by digging horizontally into the wall of the excavation or vertically just outside the hole. If you find the seepage is from a single source, you can use a collection box or pit filled with clean, coarse, gravel from ¾ inch to 3 inches in diameter. (If necessary, wash the gravel.) Drain the box or pit using a relief pipe that flows by gravity to a safe disposal location such as a "daylight" outlet (one that discharges above ground), a storm drain, or a sump pit. If the pipe daylights, it should be well-protected against construction impacts and accidentally being covered. After completing construction, you should mark the pipe outlet and warn your customer to keep it open.

### Underground Interceptor

If a sole source cannot be found, you will need to build an underground *interceptor* to collect and dispose of the inflows before they reach the foundation. An interceptor consists of a narrow ditch about 1 or 2 feet wide and deep enough to collect the lowest water course seeping or flowing laterally through the soil toward the excavation (Figure 3-1). (Occasionally, water will flow out of the sides of a ditch at different levels. The ditch must catch and dispose of all of the water that enters it from the top to the bottom of the foundation.)

If the soil is relatively stable under the moving water, line the inflow side wall of the ditch with a filter fabric. If the soil is *quick* (easily suspended in water), skip the filter fabric because it will clog with soil *fines* within a few years. Use a *waterproofing* membrane on the face of the opposite wall and underneath the gravel at the bottom of the ditch. The membrane prevents water from seeping out of the ditch toward the foundation before it rises high enough to flow out the pipe. You should fill the ditch with clean, coarse gravel (as described above) high enough to collect the highest seepage or flow. The perforated pipe should collect and convey the intercepted water to a safe outlet, such as a gutter in the street, stormwater inlet, or natural water course. Flexible corrugated pipe is easier to work with than rigid pipe, but the smooth walls stay cleaner.

The height of the pipe in the ditch depends upon the elevation of the outlet. Except in the worst situation in which the site has little or no slope, you would use gravity rather than a pump to drain the system. If you set the outlet as low as feasible and slope the pipe a minimum of 1 percent between the ditch and the outlet, you can set the pipe relatively low in the ditch. But even if the pipe must be placed high in the ditch to maintain a slope of 1 percent to the outfall, as the water collects in the ditch it will rise until it reaches the holes in the pipe and begins to flow out the pipe.

## Water Table

For water issuing from a water table disturbed during excavation, a *dewatering* system should be installed either inside or outside the building. If you only install one—all that is needed in most cases—it should go on the inside of the building for two reasons: (a) an inside system is not subjected to the heavy, perhaps crushing, load from deep backfill nor the construction debris and trash that the backfill should not contain but often does, and (b) an inside system stays cleaner over the years because it is not subjected to the water-borne sediment in the backfill *percolating* down into the gravel and pipe.

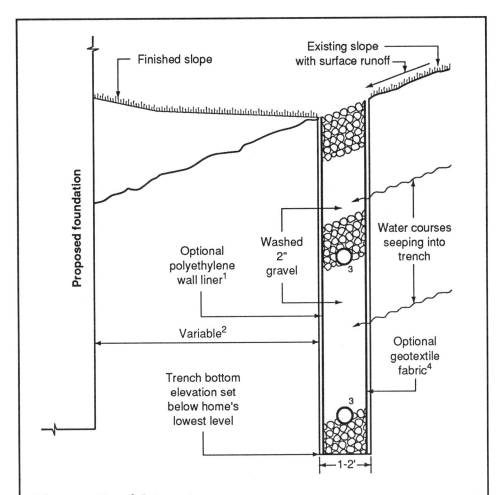

*Figure 3-1. Trench Interceptor*

1. If the soil down the trench wall nearest to the house appears susceptible to infiltration by water collecting in the trench, line that wall with 6-mil or thicker plastic membrane to prevent such leakage that might attack the house foundation.

2. The absolute minimum distance between the foundation wall and the trench is determined by where the finished slope of the future backfill will intersect the existing slope. The distance should not be less than 10 feet in most cases.

3. The 4-inch minimum diameter, perforated polyvinyl chloride (PVC) pipe should be located at the lowest position in the trench that will permit a positive outfall to an adequate outlet. If the slope throughout the length of the pipe does not average a minimum 1 percent grade, you should place the pipe at a higher location in the trench to achieve that grade. Even if the pipe is relatively high in the trench and a lot of water collects below it, the water should rise in the gravel to where it enters and flows through the holes into the pipe and out the end of it. In this situation use the plastic membrane cited in note 1 above.

4. If the soil down the trench wall opposite the house is a predominantly clayey material, use a geotextile fabric to line that wall. It will provide the first line of filtration defense against the fines leaching out of the soil. But if the soil is predominantly silty or fine sandy material, do not use a filter fabric since it may become clogged in just a few years.

5. The top of the trench shows gravel at the surface, instead of being covered with soil and grass or plants, to promote flushing of the gravel and the pipe every few years.

**Dewatering System**

A perimeter dewatering system at the lowest level of the structure appears in Figure 3-2. On the inside of concrete masonry walls, pick a hole into each void of each block as a bleeder just below the slab to drain out any water that has accumulated or might accumulate in the future. Do not penetrate the outside faces of the blocks with the pick point. Spread washed gravel (1- to 2-inches in diameter) in a layer 2 to 3 inches deep and at least 12 inches wide on top of the footing or alongside the footing all around the perimeter. On top of the gravel lay a perforated pipe at least 4-inches in diameter. For soils with a binder such as *clay* use corrugated pipe, but *quick* soils require smooth-walled pipe. Over the pipe spread another layer of the same gravel 2 to 3 inches deep to form an envelope around the pipe. Over the gravel lay a plastic film, filter fabric, or roofing felt wide enough to cover the full width of the gravel.

Inside the home, lead the pipe into a flue liner or a plastic sump pump pit located (a) where its outlet pipe can exit the building at or toward the low end of the lot so that water does not recirculate back to the pit through the backfill, (b) where it will not interfere with any future rooms, and (c) separate from utilities, laundry equipment, and the like because the pit must be accessible.

Open the holes in the sides of the pit before standing it on a bed of gravel. Surround the pit with a 3- to 4-inch envelope of gravel up to the underside of the basement slab or the underside of the sheet membrane in a crawl space. In both cases this gravel joins the gravel covering the subgrade in a basement or a crawl space that also serves as a water collection device. At the bottom of the pit embed a couple of bricks horizontally in the gravel to support a sump pump. This elevating device allows space for some of the sediment that accumulates in drainage water to settle out. Warn your customer to clean out this sediment when it rises close to the bottom of the pump. Otherwise, the sediment may block the pump intake and cause the pump motor to burn out. You should also warn the customer that cleaning out this sediment is often a muddy mess.

Installations that are subject to power outages may need a battery-backup system. If so, during construction you should rough it in and include whatever additional equipment the homeowner will need to handle the water.

On sites with loose soils (such as sands and nearly pure silts) or where the foundation will penetrate a water table, the sump pit may need to be much larger than normal, or you may need more than one pit and pump to reduce the frequency of removing accumulated sediments.

If the dewatering system is installed outside, lead the pipe around the perimeter until you can daylight it or until it is opposite an appropriate place for locating the sump pit inside. Install the connecting pipe under the footing or on top of the footing into the pit. To protect a drainage system installed outside, begin backfilling by hand for the first 2 feet upward and do not compact the soil.

**Figure 3-2. A Perimeter Dewatering System Wall-and-Floor Section**

1. Grade the surface of soil away from the foundation at 1 inch per foot for 6 feet.

2. Apply a dampproofing parging at least ½ inch thick to a block wall; and cope (make a variable thickness over a curve) it thicker at the wall-footing joint. For a concrete wall, apply a chemical or bituminous coating. If water will stand against either type of wall because of groundwater or other hydrologic condition, substitute an asphalt-based membrane about 60 mils thick. Follow the manufacturer's instructions to the letter.

3. Lay a perforated 4-inch pipe preferably on top of the footing, or alongside the footing. On top of the footing is better because it reduces the risk of water seeping under the footing and weakening the subgrade. In the drawing the bottom of the gravel bed under the slab ideally stops on top of the footing. If a capillary break is needed lay it over the first course of concrete masonry units below the top of the slab elevation to force rising damp into the gravel layer. Extend the gravel to a sump pit, a storm sewer, or an adequate surface outlet.

4. Envelop the pipe in 1- to 2-inch-diameter washed gravel and spread the same size gravel 4 inches thick under the concrete slab throughout.

5. Between the subgrade and the slab lay a 6-mil plastic membrane lapped at least 12 inches at interior joints.

6. Before placing the slab install the weeps mentioned in item 7 below and spread a sand layer 1 to 2 inches thick over the plastic membrane to keep the underside of the slab moist during its initial set, and also to absorb some of the excess water leaving the slab downward. (Perhaps the latter reason will persuade the finishers not to punch holes in the plastic.) The concrete slab should be at least 4-inches thick.

7. Punch or pick holes into the cells of blocks, and optionally install wicks, to drain to the gravel any water that builds up inside the wall. Such water may have entered the wall from above grade, penetrated the parging, or risen through or around the footing. If a capillary break is needed, lay it over the first course concrete masonry units below the top of the slab elevation to force rising damp into the gravel layer.

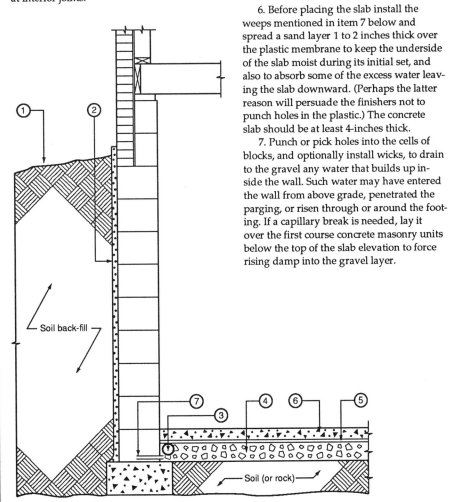

Soil back-fill

Soil (or rock)

### "Wet Feet"

Where a new house, or an addition, will have "wet feet" from the beginning, some builders install inside and outside piping systems and drain both to a sump pit inside the building. Only a few houses require double systems. However, if you do need to install two of them, they should be completely separate so each can act as backup for the other. If feasible, each system would have its own gravity drainage outfalls or have separate pumps.

An example of houses with wet feet involves a group of building sites with foundations for the basements penetrating the local water table. On one of these sites—not the lowest elevation—as the front-end loader dug deeper into a silt loam the cut became wetter and wetter. In a matter of moments the loader became mired and began to sink. The operator turned off the engine and jumped to safety. When the motion and vibration stopped, the machine stopped sinking. Later another machine hauled out the first one, and a smaller, lighter loader was substituted. The lighter machine finished the excavation with its tracks sunk to their tops as it moved around. Once this site was dewatered by a piping system and a gravity outlet, the soil was strong enough to support concrete piers that in turn carried formed footings. From there up the construction was normal, but the house sold at cost. The builder considered himself lucky not to have lost money.

## Underground Pressure

Water rising from the *subgrade* inside the house, apparently under pressure, often bubbles or roils. Before taking further action, have the water tested for chlorine to be sure the water is not from a plumbing source. If the water is not from a plumbing leak, try to locate the individual sources and collect them with individual sump pits from which outlet pipes flow by gravity to a safe disposal outside the building. If necessary, pump the water out. For extra protection, install an inside perimeter dewatering system as described above. It also would handle overflows and backups from the individual sump pits and their pipes.

If the source of water cannot be found in a case of a house with "wet feet," you should install a dewatering system before pouring the basement slab. Unless a house is in a dry area that never has had water problems the safest approach is to install a dewatering system in all basements and crawl spaces.

# Treatments

## Dampproofing

Even in dry areas, you should *dampproof* foundation walls on the outside (positive side) to protect them against overflowing gutters and windblown rain that flows down the building face. The historical method for concrete masonry walls uses a cementitious stucco treatment (often called *parging, pargeting,* or metallic waterproofing. However, it is only dampproofing not waterproofing, as discussed below). The two formulas for these dampproofing compounds materials make a grout and a stucco.

### Grout

Combine one part Portland cement (type I) with one part sand by volume. For each bag of cement, add about 15 pounds of ironite (metal shavings) to the dry mixture. Add only enough water to get brush consistency (like pancake mix). Brush the grout onto the cleaned footings and the first course of block.

### Stucco

Combine one part Portland cement (type I) to two parts sand and add about 15 pounds of ironite for each bag of cement. Add only enough water to create a troweling consistency. Apply the stucco (at least half an inch thick) over the semi-dry grout and over cleaned and patched block walls.

A couple hours after the parging is completed, rub over it with a wet sponge to prolong the curing process.

Even though parging shrinks during curing, it should not crack over the years. The ironite keeps oxidizing and expanding within the parging to resist minor structural movement, so small cracks will reclose over several months. Major structural cracks, however, cannot be bridged by parging. Because these cracks usually penetrate the wall's full thickness, they are an opening for any water, standing or flowing, to enter and spread. Water takes the path of least resistance from its present position to a new one.

When used for dampproofing, an adequate layer of parging does not require a bituminous coating although many builders still apply it. While a bituminous coating may keep water out of structural cracks for a few years, ultimately it deteriorates and becomes useless.

Parging should not let in water that percolates through the backfill unless it develops *hydrostatic* pressure against the wall. Such pressure occurs when water starts rising in the backfill faster than it can flow away through the surrounding soil. (Water's hydrostatic pressure (sometimes called head) exerts 62.4 pounds per square foot against any surface for every foot of the water's depth. For example, at its base 2 feet of standing water exerts almost 125 pounds of pressure down and sideways against every square foot of surface that contains it.)

Concrete foundation walls, whether plain or reinforced, are dampproofed differently from masonry walls because concrete inherently is less *pervious* (permeable) than concrete masonry block walls. Remember to repair surface defects and tie-wire holes and seal them before you spray on a *bituminous coating*.

## Waterproofing

You should use waterproofing where water is expected to stand against the foundation walls or under houses. Waterproofing should resist the penetration of walls and slabs by water in a hydrostatic condition. Parging alone may hold out pressurized water for awhile, but it cannot continually resist delamination within itself or separation from the wall it covers. Therefore hydrostatic conditions demand different materials. Current products include liquid and sheet membranes, fabrics, drainage cores and boards, and waterstops. (The product list grows annually so you should see your supplier when the need arises.)

## Membranes

If a house foundation must serve like a boat's skin, you must install a multi-coated liquid or thick sheet membrane. Your decision should be based on the frequency and the amounts of water to be protected against, the local availability of various materials, and prices of course. In considering price remember that correcting a poor job after it is done and covered is more expensive and time consuming than doing a more costly job that works the first time.

A multicoated liquid membrane applied over parging ought to suffice for infrequent and low-pressure ponds standing against walls. Without parging or if the foundation may be subjected to frequent or high-pressure ponds, you should install a thick sheet membrane.

Either membrane should be installed according to its manufacturer's instructions with no deviations. They are usually quite specific as to cleanliness, priming, compatible materials, lapping requirements, curing times, protection, and the like.

## Boards and Geocomposites

You may need to use a *protection board* (a ½- to 1-inch thick panel attached to the wall) or a combination protection board and vertical drainage core (*geocomposite*) because (a) abrasive material in the backfill might tear the membrane and allow water to penetrate, (b) some membrane seams may open, or (c) the membranes may peel off the wall if they are subjected to prolonged standing water. Thus the situation may require a drainage board that combines the functions of protection with water conveyance.

Protection boards may be made of expanded or extruded polystyrene (that also serves as insulation), bituminous fiberboard, and other products. Geocomposites include, but are not limited to, expanded polystyrene beads bound with bitumen, nylon matting laminated with filter fabric, and waffle-like drainage core laminated with filter fabric. Drainage board requires a piping system at the base of the wall to collect and convey away water that accumulates in the backfill. This system resembles the inside dewatering system already described.

## Waterstop

Another device called a *waterstop* is sometimes used in concrete walls at construction joints. (A construction joint is a deliberate separation of panels. Builders and remodelers use these joints when they must stop the concrete work and resume it later because of shortages of time or materials or when design or sound practice calls for an *expansion joint* designed to    allow for movement when concrete expands from heat or shrinks from cold.) Current waterstops are rubber, bituminous, or plastic premolded fillers, some of which include clay materials that are placed against one side of a joint and then compressed tightly by the placement of the next panel. While in theory they should prevent the passage of water from one side of the wall to the other, in practice they often leak. Before you use any type, check on others' experience with a particular brand. In addition you should dampproof or waterproof joints in concrete walls by methods previously described.

# 4

# *Treatment of Leakage After Construction*

If a customer calls with a residential water problem, you need to patiently and carefully study the problem before starting attempts to fix it. You should look for simple solutions to simple problems because unnecessarily complicating a situation wastes resources—money, time, and effort. If you use the correct diagnostic procedures in the beginning, you probably will be able to solve the problem the first time.

You need to resist the temptation to come to the scene with a solution based on the symptoms the customer described over the phone. This temptation is especially strong for builders who have successfully treated several production houses with the same problem. If the problem persists and the customer continues to complain, a troubleshooter may be tempted to make a cursory inspection and continue to treat the symptoms based on previous experience in similar situations. With luck the second or third effort solves the problem, but most people would prefer to be do it right the first time. Before determining what procedure to use in a given situation, you need to look at the concepts of causes, mechanisms, symptoms, and effects.

This philosophical exercise is necessary so troubleshooters will not just treat symptoms and neglect to look for—and of course not find—the mechanism of the problem. Of course you should try to find the cause by working backward from the mechanism, but sometimes you cannot, such as in hidden groundwater situations.

Cause is used here as the why of an occurrence. Mechanism refers to how it occurs. In the example of water leaking out the bottom of a steel gutter through rust holes, the leaks are the symptom or the effect, the holes are the mechanism, and the water running into the gutter from the roof is the cause.

The first example should be clear, but other cases are more complicated, for example, treating a wet basement during dry conditions. You cannot observe natural roof runoff to determine if gutters and downspouts are working (although a water test using a garden hose might be worth a try), but you see where water has ponded on a flat planter area extending 5 feet from the foundation wall. "Aha!," you may think. "The solution is to remove the shrubs, add enough proper backfill to reestablish the correct positive grade, and replant the shrubs." You could have the work done and go on to the next task. (Proper backfill is discussed below.)

After the next thunderstorm your customer may call to report his or her wet basement is improved but still wet. Upon another inspection you might see holes in the newly replaced backfill where water has infiltrated (penetrated) and *percolated* (migrated). In the basement you again find dirty water standing on the floor. After further discussion, your customer remembers that water was sheeting over the gutter during the strong winds accompanying the previous rainstorm. This case has only one cause, but it has at least two mechanisms, and you treated only one, the inadequate grading. The gutter-downspout problem, whatever it is, was not treated. This problem also may have a third mechanism that only becomes apparent after another look-see (see Diagnostic Procedure below).

Sometimes the homeowner will ask you to coat the walls on the inside with a "waterproof" paint or seal them with a high-technology liquid membrane in the hopes that water will not penetrate into the home. The answer is that such treatments applied to the negative side of walls is primarily cosmetic. If the problem is minor it may involve only enough water to dampen the walls but not flow onto the floor. However, if the problem is worse than minor, the water must be kept out of the wall. If the wall is built of concrete masonry units (CMUs), this defense may be at the expense of the wall's integrity because water standing in the cells of blocks over a period of years may dissolve the mortar. The preference should be to control the water on the outside as discussed below.

# Diagnostic Procedure

Each situation will vary in complexity and in the time you will need to explore it. You will learn how much time to devote to a particular problem from the practice of diagnosis and from the results of the treatments. If you are to err, let it be on the side of thoroughness. Customers often have not recorded the weather conditions when an incident occurred. And they may not know, for example, whether water came through the basement wall during or shortly after rainfall or many hours after the rain ended.

Start by exploring the problems with the homeowners. Let them show you what and where things occurred in the sequences they choose. Try to keep them from jumping back and forth from area to area or room to room, but otherwise let them set the pace and direction. Where the problem seems complex ask for a written summary of events and make some notes yourself. Using a pocket recorder saves time and prevents forgetting details that might be important. Be thinking about any details that are unclear. Ask about them now and review them when you reinspect conditions later. Eliminate plumbing and hydronic heating leaks before proceeding further.

Most visible water flows downhill so check the highest area or level of construction first. Maybe you cannot get to the top of the chimney on the first visit, but if the symptoms point that way (whether in the attic, the basement, or in between), you may need to see it soon. Look closely at all the areas the homeowner showed you, plus any adjacent locations that may be related now or in the future. At least cursorily look everywhere else too on the off-chance something unrelated appears.

At this time you need to keep an open mind even though you might be tempted to draw conclusions and rush to a solution. After finishing with a homeowner, spend time alone or with one of your employees going over everything both physically and mentally.

After the searching comes thinking. First, try out theories connecting symptoms with mechanisms. If connections do not fit, do not force them. You may discover another explanation later. You may have missed something, or you may have to requestion the homeowner or talk to the neighbors. After discovering all the mechanisms, think backwards to one or more causes that might be treatable. Devise a plan that includes the treatments in the proper sequence.

You may not find the cause because to excavate or demolish certain areas may not be practical or feasible. In that case you will have to treat the mechanisms in an organized manner. And occasionally you will treat symptoms because you cannot discover the mechanisms or causes. This approach will waste some time and money if you must try more than one major solution.

As an example, consider the dining room windows at the gable end of a two-story brick home. The owner points out the water stains on the plaster or drywall above, below, and between the frames. He or she cannot remember exactly when it occurs, but the customer knows the leaks do not occur during every rainstorm. In your search outside you note the tall gable faces the predominant wind direction during storms. You observe a few holes in the brickwork, but apparently not enough to cause the amount of damage you saw inside. The angle-irons (steel lintels) over the windows have their horizontal legs caulked closed instead of being left open. This prevents water that may have accumulated in the wall from weeping out over the window heads, so you search higher looking for points of entry from blowing rain. None appear in the brickwork, but there are gaps between the bottom of the roofing slates and the rake boards. Without more symptoms, you recommend that the customer have a roofer install a rake flashing under the slates overhang at the gable end. You also ask for feedback about whether the "solution" works.

The previous example illustrates the difficulty with some water problems that have obvious symptoms but present little information on how they occur, even when someone is there to watch it happen.

# Treatments for Leaks

As home inspectors will tell you, residential water problems predominate at the roof and the basement. Note how one area is at the top of the house, and the other is at the bottom. Starting with the top, you need to look at symptoms and mechanisms, their respective causes, and how to treat them.

## *Chimneys*

Leaks at ceilings and walls are a common problem. The following items may have separated, opened, torn, or disappeared from a masonry or metal chimney:

- caps and *washes* (the concrete cover surrounding the flue)

- storm collars
- base and counter (step) flashings
- mortar joints
- rakes at *drop-offs* (where the chimney size is reduced above the breast)

All of these items should be closely inspected for obvious problems and corrected by one or more of the following procedures:

- Narrow cracks (up to ¼-inch wide) at masonry caps or washes should be sealed with the best quality sealant (caulking).
- Wider cracks often need more widening, enough to install a concrete (not mortar) filler. When the filler has dried sufficiently, it should be sealed at each joint.
- Retighten or replace storm collars on metal chimneys in the rare event that they get loose and slide down, or they get bumped and bent.
- Retack and reseal loose or leaking chimney flashings. Damaged and worn flashings may be too thin to withstand wind action and hold sealant, so replace them. Both types of chimney flashings (counters and bases) are fastened to the chimney, but they are separate from each other. In extremely old chimneys the flashings were embedded in the mortar joints and seldom worked loose even if some mortar fell out. Modern flashings are often nailed into the mortar and sealed tight with a roofing cement (mastic). Flashings that are only nail-tacked and sealed into masonry can work loose and expose the chimney to surface runoff. No matter how a flashing is installed it may tear from impacts such as from tree limbs, but usually the step metal (overhanging the base metal) bears the brunt. To prove that a flashing is leaking, you may need to spray water from a garden hose on it and adjacent materials.
- If mortar has fallen out of joints or if old cracks appear through or across the joints, sealant will fill small holes and narrow openings. Wide gaps need replacement mortar. If these symptoms appear in new homes, a structural defect may be the mechanism, and you should determine its cause. Sealing is the proper, if temporary, treatment to keep out the weather.
- Leaks into chimneys from the top can be treated in two ways. One is to find the source of the water, such as a downsloping tree limb from which runoff cascades down into the chimney. In contrast the water dripping off some leaves usually does not create much of a drainage problem down the chimney and into the smoke chamber. The second treatment involves installing a metal rain cap on top of the open chimney cap. Its roof sheds water, of course, and the screen surrounding the metal framework keeps animals and birds from nesting on top of the fireplace damper.

Although this book is not about structural problems, a short discussion should help you. A masonry chimney is obviously quite heavy and may settle more than the wall connected to it. Just a fraction of an inch is enough to cause a separation between two similar or dissimilar materials because they "feel" different stresses. If the settlement is short-lived, say a few months, the repair can be a symptomatic one. Thus the sealant will suffice. But if the settlement continues for several months or more, probably the underlying soils condition will have to be treated before you treat the cracks.

Where a chimney width is reduced (for example, above a fireplace), a rake occurs. To shed water it should be sloped and covered with a relatively impermeable cap such as flagstone. If the cap falls off or separates from the brickwork, replace it or seal around it.

## Roofs and Roofing

Leaks through the field of a roof covering are rare except when it is nearly worn out. Then they occur where—

- individual pieces have blown or been torn away
- corners have broken off
- edges have separated
- holes have developed
- layers have delaminated and so on

Of course, the foregoing problems also can occur in roofing with substantial life remaining. These problems are obvious when you view them from the roof and maybe even from the ground with the aid of binoculars. Treatments are also obvious.

*Crickets* are small A-roof sections that divert the runoff from above chimneys away from the longer side wall of chimneys. Without them a heavy downpour or drifted snow could overwhelm the regular flashing between the chimney and the roofing. Whenever a shed roof ends at a chimney wall that is more than about 3 feet long be sure a cricket is behind the chimney.

Sometimes harder to discover are defects at roof penetrations such as ventilators, skylights, plumbing vents and stacks, heating and cooling trunk lines, cables and tubing, equipment supports and legs, and dormers. At these penetrations metal and nonmetal flashings include variations of the bases and counters discussed above under Chimneys, round and rectangular sleeves, collars, flanges, and boxes; *pitch pockets* (now out of fashion); and rectangular pans. Defects also occur at ridge caps that cover the peaks and upper shed edges of roofs and at copings that cover the tops of walls and railings.

The flashings at penetrations and covers leave gaps for the passage of water when they tear, deteriorate and decay, loosen, sag and slump, and separate. Among the causes are wind, structural movement, solar radiation, and cooking oil residues that escaped from the kitchen via a power ventilator. The simplest treatment for a short tear, gap, or slight looseness in an otherwise adequate flashing is to apply the proper sealant. More advanced problems such as deterioration, flapping in the wind, and slumps may require repair or replacement.

Inspecting low-sloped and flat roofs during a rain may reveal the leak location, but often the origin of the leak is not directly over the evidence of it. If the attic is accessible you may more easily find the sources by following the stain path left by the water. Water testing a roof with a garden hose may be useful. Drive the stream against penetrations as a simulation of wind effects, starting at lower elevations and working upward. Otherwise, testing from the top down might let higher mechanisms obscure lower ones. Similarly, flooding flat roof field areas by sections may reveal a leak; but one test area should dry before you move onto another unless they have a water barrier between them.

Leaks at dormers and under mechanical equipment that sits directly on the roof (instead of on an elevated platform) can be especially hard to find. Dormer flashings are so extensive that minute holes or gaps can escape all but time-consuming examinations and water tests. Likewise, if you cannot see under a rooftop condenser that has a grille on top, you should spray water down inside to test the bottom pan's integrity.

## Gutters, Downspouts, and Scuppers

Ice dams over gutters and downspouts should be obvious, but blockages by litter are less so. In wooded areas cleaning and flushing twice a year is not too often.

New metal workmanship should not leak, especially seamless extrusions, but also soldering or riveting and sealing. Leaks occur in older installations when solder joints open, galvanized steel rusts through, or downspout seams pop open from blocked water that freezes.

The placement of the gutters may create problems for new and old work. For example, gutters may—

- slope away from the downspouts versus toward them or slope back toward the house
- be too far under the roofing overhang
- be too small for the roofing material
- be too low on the gutter board (fascia)
- attach too loosely to the building
- be improperly connected to the downspout
- allow water to cascade over them at the feet of valleys

For appearances, gutters should slope only up to 1 percent; but because their slope is usually determined by the top edge of the gutter board (fascia), an out-of-level gutter board may result in a backsloping gutter.

At eaves the roof sheathing should overhang the fascia from ¾ to 1 inch, and the roof covering should overhang the sheathing about the same amount. (Stopping the sheathing behind the fascia theoretically sounds neater, but practically may leave a gap behind the fascia which the roofing felt does not bridge, thus leaving a gap for water entry.) The roof covering should overhang the gutter up to a third of the width of the gutter. This overhang serves as a lead-in device to the gutter underneath.

When the roofing overhang is too short, water may curl back between the shingles and the sheathing, or it may curl under the sheathing. On homes with a roof overhang, this curling action may cause the decay of the sheathing and some pieces of the cornice, such as the soffit. In homes with flush eaves, the decay may be accompanied by leakage into the building. You can cure the too-short overhang by installing flat metal under the roofing beginning as a lead-in flashing over the gutter and up the slope past the starter course.

If the roof covering overhang is so long that it covers more than the back third of the gutter, water may spill over the front of the gutter in certain conditions. For steeply pitched roofs both slow and fast runoffs may still drive into the gutter. But for shallow pitches the fast runoffs overshoot the gutter. You can cure the second by padding out the fascia enough to place the gutter in the correct position, or you can cut off the excess overhang.

Gutters should be as tight to the underside of the roofing as possible. When standing out in the yard on a spot to be determined in each case, you should not be able to see over the front edge of the gutter (except at and near the downspout ends) and look up the plane of the roofing. Wherever you can see the roofing in-plane, hard rains will be likely to overshoot the gutter in those places. Some of you who work in heavy snow areas will protest that this advice is a recipe for gutter tear-offs when the snow slides down the roof. The answer is to use plenty of snow guards near the eaves and up the slope at least a few feet to break-up the slides.

A 5-inch-wide gutter probably is too narrow for heavy wood shakes. Even when the minimum overhang occurs, the thickness of the shakes puts their upper surface up so high that heavy flows off a shallow pitch may overshoot a narrow gutter. The extra inch of a 6-inch gutter is inexpensive insurance.

When the gutter is not tight to the fascia, you can look up and see light between them. Slow runoffs may flow through the gap and run down the face of the building or just land on the backfill close to the foundation. In these situations, tighten the gutter according to the correct procedure for the hanging system in use.

An inadequate connection between a gutter and a downspout that allows water to escape the path may not be apparent in dry conditions. But when a loose downspout slides off the elbow (gooseneck) or comes apart elsewhere, you should see it from the ground.

Scuppers may be cracked between the roof where they receive runoff and their outlets beyond the fascia or frieze. Escaping water can enter the top of the wall of the building and run down by gravity until it comes out on the inside somewhere.

From a distance most connection problems are obviously hard to find. The best approach is to look for them during rainfall or snow melt or test for them with a hose.

During heavy rains, roofs with low to medium pitches may send runoff over the gutter intersections at the feet of valleys. To prevent this attach diverters to the outer edge of the gutter. These will be 2 to 3 inches tall and extend about 1-foot each way from the intersection. They should intercept the water flowing down the valley and direct it into one or both of the gutter legs as discussed in Chapter 1.

Another source of leakage that appears on the ceilings and walls (usually at their intersections under eaves) is ice dams. Obviously they occur in cold climates during the winter. Prevention usually is easier than correction, so you should follow these methods during construction:

- Lay a 3-foot wide geomembrane under the roofing beginning at the ends of the eaves, as discussed in Chapter 1.
- Eliminate gaps in the attic insulation at or near where the rafters and ceiling joists intersect. You want to prevent the heat from conditioned spaces from escaping and warming the roofing at the eaves (the cold-roof method shown in Figures 4-1 and 4-2).

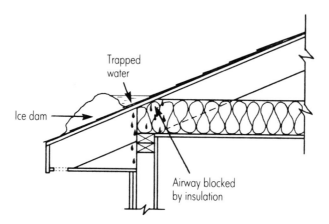

**Figure 4-1. Ice Dam**

*Thick insulation blocks the airflow, allows heat to build up, and create a warm spot on the underside of the roof that will melt snow on the roof. The melted snow runs down and refreezes on the cold part of the roof to create an ice dam.*

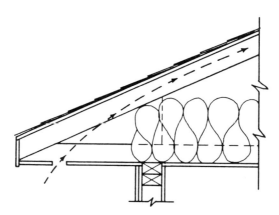

**Figure 4-2. Prevention of Ice Dams**

*You can use a cantilevered truss construction to provide a passageway for ventilation air from soffit vents where thick insulation would otherwise interfere. This construction serves to "air wash" the rafter bays and keep the roof temperature even.*

On an existing home you might try fastening electric heat cables (according to the manufacturer's instructions) on the roofing where ice previously formed. This method is not a permanent solution, however, because the cables wear out or could start a fire if the insulation is melted.

If a house has no gutters it should have exposed gravel trench drains centered under the drip line of the eaves. Depending on climatic conditions, the trenches should be 12- to 18-inches wide. They should be at least 6-inches deep and contain gravel from ¾ to 2 inches in diameter. If the roof runoff will be voluminous make the trenches deep enough to also include perforated pipe 3 to 4 inches in diameter. The trench drains should outlet to daylight or to storm drainage. You should warn the homeowner to flush the drains periodically.

## *Exterior Walls*

*Skin* is an all-inclusive name for the outside walls of a structure. Rare are the leaks directly through the fields of masonry and frame. Instead most leaks occur at skin penetrations, such as around fenestration (windows and doors), *wall jacks* (the special vents with doors such as those used for bath and kitchen fans, wires, vents, and pipes).

Most bricks and stones are waterproof, except during the highest winds. Even mortar and concrete joints do not leak except where microscopic-size and larger holes and gaps occur in the joints or at their edges. If you can find such openings use a high-quality, exterior, flexible sealant (caulking) to plug the holes and gaps. If you prefer a coating, use one that resists water moving from the outside surface inward, but allows water vapor to escape from the inside of the building outward. Do not use the silicone-type coatings that prevent liquid and vapor from moving through the masonry in either direction.

If siding is backed-up by sheathing, subsiding, or other barrier, the siding is unlikely to leak directly to the inside—not even through splits, broken corners, and other defects. But water that penetrates the siding may travel down short distances to gaps in the back-up, or down long distances to foundation wall plates. In both cases, it can find one or more ways inside the building.

Leaks around and through penetrations mostly result from inadequate sealing and flashing, such as—

- openings between masonry chimneys and frame walls
- loose bathroom and dryer vents whose flanges have gapped away from the wall surface
- separations between entrance features and adjoining walls
- electrical service sleeves with sealant missing around them or around the cables
- gaps under sliding glass door thresholds
- unsealed joints between window trim and abutting siding
- split sills at doors and windows.

Some moisture control specialists now believe that blowing precipitation leaks through a building's skin regardless of your waterproofing efforts.

They want to transfer their experience with high-rise buildings, subject to extreme winds, to one- and two-family dwellings that do not exceed 40 feet in height. Their solution is to let the water penetrate the skin but stop it at the next line of defense such as a tight sheathing or barrier. From there it is supposed to travel down to through-wall flashings that weep it back outside to a safe outlet. This approach is based on the *rain screen* principle of (a) equalizing the air pressure on both sides of the leaky skin and (b) letting it increase at the surface of the back-up surface that is supposed to be weather tight.

On paper the rain screen looks like an effective device for residential construction. In practice the as-built rain screen may resemble the drawing but not fully agree with it. Brick veneer over framework is frequently cited as an exceptionally clear example. In the drawing the air space between the back of the brickwork and the framework is open and clear. On the job the back of the brickwork is barely an inch from the sheathing. And in contrast to the drawing, that narrow air space is punctuated by code-required wall ties. Furthermore, as the bricks are laid some mortar is squeezed out the rear of the joints, falls on the wall ties, and otherwise builds up in the air space that has become somewhat of a *collar joint*. A collar joint is the vertical space between the back of brick facing and the adjacent back-up masonry that is filled with mortar.

Allowing leaks through the veneer's head and bed joints into a partially blocked air space traps water standing against the sheathing or other barrier that at best is a water retarder, not a waterproof membrane. In addition a waterproof membrane is undesirable because it prevents water vapor from migrating from inside the dwelling to the outside during cold weather (see Condensation below). If the trapped water develops enough hydrostatic head, it is likely to find a pathway to the inside.

Widening the air space would leave more room for fallen mortar, but it still does not address the primary problem of poor workmanship (leaky joints) on the outside. The rain screen principle in this instance treats only symptoms not causes and mechanisms.

The conclusion here, at least for brick and stone veneering, is that the primary defense against leakage into a home is a weather-tight skin, and the secondary defense is a weather-tight backup. Relying only on a weather-tight backup is risky.

All-frame construction can be a different matter because a clear air space between siding and sheathing—open to *ambience* (the outside atmosphere) at the top and bottom—is practical construction. All-frame construction is also appropriate for dissipating water vapor that has migrated from inside the home.

Leaks at the skin penetrations listed above usually can be stopped by improving the weather tightness around the penetration. Flashings are more permanent and rely only secondarily on sealant, but in some cases they were never installed. To install many of them after the fact is often prohibitively expensive and may be unnecessary. Sealant is less permanent, but because trimwork needs periodic refinishing or repainting even on brick homes, the sealing can be touched-up or redone when the painter revisits. The exception to adding more sealant to a leaking joint is when it will do no good. Then the forgotten or excluded flashing will have to be installed no matter what the cost. For example, when a swinging entry door leaks above the threshold after the threshold has been replaced, try attaching a water table that overhangs the threshold (Figure 1-1).

When the leak is through, rather than around, the fenestration often a fitting problem is the cause. Adjust or replace loose windows so that they are nearly airtight not just watertight. Adjust a wall jack door that does not properly close. Seal around the cables inside an electric service sleeve.

The reminder here is to look for—and correct—deficiencies even at unlikely locations.

## *Foundations*

### Surface Treatments

Evidence of leaks in this category occur at walls and floors of basements and crawl spaces. How can you tell whether the penetration is (a) high on the wall with the mechanism being runoff at grade backing into the house, (b) a subsurface penetration low on the wall at the wall-floor joint, or (c) somewhere in between through cracks in the dampproofing? One way to distinguish between mechanisms is to check for the typical triangle-shaped stain or wet area on the wall.

If the triangle is tall with the apex close to or opposite the grade line outside, the penetration probably is high on the wall and spreading laterally as it travels downward. Treatment in this situation involves correcting one or more downspouts and perhaps correcting the grading next to the house. A downspout improvement may require raising or lowering a downspout shoe, pointing it in a different direction, or extending it farther out into the yard. Surface runoff is preferable because future problems can be seen if someone looks. For example, if the downspout dumps on a patio with too little slope or backfalls toward the home (Figure 4-3), the downspout may have to be relocated or eliminated. Eliminating a downspout usually requires increasing the size of the remaining spout at the other end of the gutter.

Another problem could be a downspout that was left in place while a deck floor was laid around it or a downspout caught between the main house and a new wing. Are the outflows from these spouts trapped whereby they infiltrate the foundation backfill? If so, they need extensions although access can be difficult. Or if access is now impossible, relocate or eliminate the downspout as noted above.

Along with downspouts that dump alongside homes, backsloped or buried splashblocks, sump pump outlets, and condensate drains from central air-conditioning also cause problems. A sump pump is especially troublesome because of its large discharge. Even when directed onto a long splashblock, this flow can quickly erode the ground surface, create a soil dam, and perhaps even start flowing back toward and then down the foundation wall. You should pipe it a long distance.

A condensate drain is less of a problem because of its small discharge, but over a period of years it can dump enough water alongside a window well, for example, to keep a basement wall wet all summer. In such a case,

*Figure 4-3. Old-Patio Problems*

*This broken and settled patio both ponds water and allows infiltration at the gaps between individual flagstones. The splashblock serves a hidden downspout that should be eliminated and replaced by doubling the size of the downspout at the far left.*

add an extension pipe to the drain.

Since the backfill alongside a foundation continues to settle for at least a year or more after construction, it may need to be raised to reestablish the original minimum slope of 1 inch per foot for 6 feet or so out from the foundation. If the yard beyond the backfill area also needs regrading the minimum slope should be 2 feet in 100 feet (2 percent).

A cautionary note regards frequently heard and read advice on sealing new backfill with a layer of heavy clay. It might be good advice to follow if you are sure the basement leakage is caused by roof and surface runoff from this property only. But if a possible subsurface water component comes from this or other yards, sealing the backfill with clay and/or a plastic cover will prevent the evaporation upward of water standing at the base of the foundation. In some instances, water that has accumulated around a foundation cannot penetrate native soils in response to gravity. Therefore the water must rise upward if it is to go anywhere (see Appendix). Figure 4-4 shows water standing in a shallow hole at the corner of a garage.

Using a brick, plastic, wood, or other edging to separate the foundation planting area from the rest of the yard is risky. It often prevents water from running away from the foundation. If it must be used, leave spaces or cut gaps between individual members.

Hard pavements, such as patios, sidewalks, and driveways, that backslope toward houses are obviously much harder to regrade than yards. Short of breaking and replacing them, you can take corrective measures. You can build a barrier that acts as a combination dam and diverter that turns water at the foot of the slope toward a safe outlet. Such a barrier

*Figure 4-4. Elevated Local Water Table*

*Water stands in a shallow hole recently excavated at the corner of a garage. The water in the hole show the elevated local water table that could lead to foundation problems if it is not treated.*

could be an earth berm, a garden tie or two, a masonry wall, or asphalt or concrete curbing. You could install a yard inlet at a low place and make it outlet by gravity to daylight.

If an original patio laid adjacent to the foundation has since been covered by a low deck, the mostly unseen patio may be broken and settled enough to permit rainwater that infiltrates the deck to pond against the foundation, and perhaps enter the backfill and penetrate the wall. Correcting this situation would require, for example, inserting a heavy plastic sheet between the bottom of the deck and the top of the patio. The sheet may have to be hung from the deck framing if it cannot be laid on the patio surface. If access is impossible from underneath the deck, various deck boards would need to be removed to permit it. This exercise shows the merit in removing patios and correctly sloping the subgrade before decks are built over them.

When the wet signs on the wall or floor near the perimeter do not form a tall triangle on the wall, diagnosis of the location mechanism, whether high or low, may be more difficult. The water may be coming in low from a subsurface source. Perhaps it is entering where the wall joins the footing and moving from there up through the floor subgrade, and then in through the wall-floor joint. Or the water may still be entering high in the backfill from a surface source but traveling down the wall "looking" for an entry place. Both sources might be at fault, but that probability is low.

Treatment of a perimeter problem usually starts with mechanisms at the highest location and progresses downward. Two reasons are experience and economy. Experientially the probability is high that runoff from the same home's roof is the culprit. However the runoff from this yard or other yards in the neighborhood toward this yard, may be overloading (surcharging) the backfill. Economically surface-and-above treatments for this condition are less expensive than the ones below grade.

This chapter has already discussed corrections to roofing and guttering to protect against water leaking into a home and to reduce its cascading or sheeting down the skin. They should be first on the treatment list. Yard drainage alternatives would come next. Chapters 2 and 3 introduced and discussed correcting grading problems before and during construction. At those times you can do more landforming because no buildings and other improvements interfere with that work. After construction, problems may be complicated by fences, new trees, gardens, playsets, swimming pools, patios, and the like. These improvements may have landlocked portions of the lot or reversed runoffs so that they flow toward buildings.

Normally you can tell by looking at a site where surface water flows from and to, except where the slope is flat or nearly so. But even at low-slope conditions, ponding usually leaves some evidence of where it goes after it rises high enough. Confirming your conclusions with the owners is helpful. But often they do not observe their yards during precipitation runoff, or they do not understand what they are seeing.

Start checking at the farthest reaches of the property to find how far upgrade the flows start and how extensive the contributors to that flow are. For example, how many roof runoffs and yard runoffs are involved? This data, when studied, gives you the sizes and locations of the devices needed to handle those flows. Often all that is needed is a small swale to collect or intercept a neighbors' runoff and convey it past or around the home being treated. Perhaps this swale also should collect some of the roof and surface runoffs on that home's yard. At the other extreme might be a rock-lined interceptor with revetments or ground covers to protect the side slopes and berm.

Listing all the different situations you might encounter on this and neighboring properties is not feasible here, but three that could challenge you are listed below. Any of the three may have substantially increased flow volumes or velocities because of the concentrated nature of roof runoffs (also see the examples in Chapter 5):

- recent construction of a separate building, such as a large garage
- a new addition to an existing house
- a major regrading of a yard during installation of a swimming pool

Such examples might require enlarging an existing drainage device, adding a new one, or combining some of the different methods and devices that were listed in Chapter 1. Combinations are almost unlimited in scope and depend only on your ingenuity and budget.

While the homeowner may be as much interested in cosmetics as performance, performance should be more important to you. You should be careful to err on the side of overdesign and overbuilding if you want the job never to bother you again. Also impress on the homeowner the need for routine maintenance.

## Below-Grade Treatments

Customers often demand subsurface drainage from—

- downspouts (with or without shoes)
- yard and patio inlets
- swimming pool apron drains
- sump pump outlets
- areaway drains and the like.

They do not want excess water running across the yard, driveway, and patio. They dislike replacing the mulch washed out of foundation planters. Safety may be a consideration, or perhaps the yard, driveway, and patio have too little fall (slope) across them for adequate surface runoff.

Subsurface drainage (underdrains) is expensive and is subject to failure from underground blockage and breakage. Because typical homeowners do not realize when underdrains are not working. In their minds they usually do not even connect a newly wet basement with an overflowing gutter, let alone an inoperative underground drain pipe. Therefore you should argue against subsurface drainage unless it is necessary or exceptionally useful. As part of the sales process, you might also sell the homeowners an annual maintenance contract to keep the subsurface drainage operating properly. This service should include cleaning and flushing the system.

Before installing underdrains, you need to decide whether perforated or unperforated pipe is better for the specific application. Perforated is intended primarily for collection from external sources, and secondarily for conveyance of the water collected in the pipe. It is the best choice for a dewatering system under a basement floor. Unperforated pipe is for conveyance only although a gravel envelope around it will act in both ways.

For a downspout extension, unperforated pipe usually is the better choice because a partial blockage might force water out the perforations of the other type. If you are also trying to dewater an adjoining wet area, start with unperforated pipe and continue it a pipe length or two beyond the building. At the end of the unperforated pipe, start with perforated and continue it to the outfall.

The gravel envelope is used mostly with perforated pipe, but occasionally with unperforated. It should surround the pipe a minimum of 3 inches. Coarse gravel (¾ to 3 inches in diameter) is better than the small sizes because of its greater permeability. The gravel envelope's purpose is two-fold: (a) to increase the capacity of the system and (b) to filter the water to reduce the erodible fines that penetrate the gravel and that subsequently are carried in the water coursing through the gravel and into the pipe (piping). For residential work the size of the gravel is a compromise between easy handling and high permeability. In a case where the underdrain will be permanently inaccessible, you should consider using a "graded" filter. This layered filter consists of small gravel next to the soil and large gravel next to the pipe. The two sizes should not be mixed together, although it is laborious not to. The large gravel's diameter should be greater than the pipe perforations.

Filter fabrics (*geotextiles*) that surround the gravel envelope must be used with caution because they can get clogged by eroding fines within several years of use. If the soil has a clayey content that acts as a binder, the filter fabric may be used. But if the soil is silty and readily erodes when exposed to moving water, the fabric probably should not be used (*May* and *probably* are the operative words here because research is still underway.)

Exterior underdrains should slope at a minimum of 1 foot in 100 feet (1 percent). This percentage should be an average amount because rock or other impediments may flatten out the slope. The slope should not reverse because of the probability of pooling and the resulting deposits of litter that ultimately will block the pipe.

Homeowners read about drywells and think they can save the price of extended underground pipes by locating the drywells close to foundation corners. The first misunderstanding stems from the size of drywells that causes them to easily and quickly overflow. For instance, a 4-feet-square hole that is 4 feet deep contains 64 cubic feet. But since a drywell is filled with gravel, broken masonry, or the like with an average porosity of 50 percent, the capacity is reduced by half to 32 cubic feet. Multiply the net capacity by 7.5 (the number of gallons in 1 cubic foot) to get 240 gallons.

If the contributing roof area is 600 square feet and 1 inch of rain falls on it, the volume of rain is 50 cubic feet. Multiplied by 7.5 that 50 cubic feet of rain translates into 375 gallons. Thus the runoff exceeds the drywell's capacity by 135 gallons. If the homeowner is lucky the drywell overflows out the top, but often instead, the drywell leaks out the bottom and sides toward the building. You have already surmised the rest of this story—a damp or wet basement.

If a drywell is the only alternative, you should meet the following requirements:

- Build it at least 20 feet from the building.
- Make it large enough to hold the rainfall from a 10-year storm for your area that lands on that portion of the roof being drained.
- Install a relief pipe that rises from middepth inside the drywell and exits below the top of the drywell. From there, extend the pipe several feet downgrade from the drywell to daylight.

After the high and surface treatments are tested by heavy rains or melting snow for at least a year and not found wanting, you have succeeded. If the basement leaks continue, even though somewhat abated, you will need to treat the subsurface mechanism.

For a relatively new home, say up to a few years old, that has isolated wet spots on the walls you should consider outside excavation to see if the walls are leaking because the dampproofing or waterproofing has failed. Patching of grouting cracks, and then sealing over them with bitumen, chemically-based products, or metallic parging should stop the water penetration. If the excavated backfill is very rough-textured, add a protection board to the wall before rebackfilling.

For an older home that has this problem in several locations on different walls, such exterior treatments are severely expensive.

The foundation plantings have to be removed and then heeled in somewhere to wait a year or more for the settlement of the new backfill before being replanted. And the job requires extensive labor to excavate, apply water-resistant materials to cleaned walls, and backfill the excavation. If the roof and yard runoff corrections do not stop the continuing penetration, use a dewatering system to do so. Treatments on the negative side of the wall, if they work, will trap water inside the wall, and perhaps lead to its ultimate deterioration.

### Dewatering System

Unless a system is already in place inside the foundation, dewatering of the basement or crawlspace of an existing home is an expensive treatment. But a system may be in place because more and more builders in areas subject to basement water problems are installing perforated plastic pipe around the perimeter walls on top of or alongside the footings during initial construction. (In older homes agricultural [terra cotta] clay tile pipes may be in place, but most installations were outside the footings and are probably blocked and/or broken by now.) Modern builders probably also installed a sump pit or two into which the plastic pipe exited. If you can find a roughed-in system, and it is usable, test it by sending water from a hose all around the pipe. If it works well, drop in a submersible pump that will exhaust to an approved drain inside or to a safe outlet outside.

Dewatering systems serve two purposes: first is the obvious interception of water penetrating the bottom of the foundation that is sent to a safe outlet by pumping or gravity. The second takes advantage of the less well-known principle of *drawdown*. The subgrade under a slab can hold a certain amount of accumulated water from wherever its source, without the water becoming visible. Many homes have such water under them all the time or just during times of high water tables or when the nearby soil is saturated. Without a dewatering system, the water that enters through the foundation and becomes visible, is the incremental amount over and above the residual amount that is invisible under the slab. A sump pump will eject most of the water that flows into its pit. Therefore, if the pit is deep enough and the subgrade is permeable enough, most of the residual water will be pumped out (drawn down) before the next storm arrives or the water table elevates. Thus the "vacated" space leaves room for incoming water that may not rise enough to become visible on the slab.

If you have to install a dewatering system, consider first trying only one of its components, the sump pit and pump. This choice would be indicated by the same single location getting wet each time. Or if more than one location gets wet, pick the location that gets wet first and dries out last. Later, if the problem areas multiply or new locations occur, install the perforated pipe and gravel under the slab along the perimeter as described in Chapter 3. Other basement water penetration symptoms might be in addition to—or separate from—the triangle discussed above:

- Water comes through the floor around penetrations such as columns and pipes or around the outside of a floor drain.
- Water stands in a low spot in the floor out away from a wall.

- The corners are wet in three or more places.
- Basement window wells fill up and flow over the sill and down the wall.
- In a situation similar to window well problems, water comes in under basement doors at the foot of areaways or it issues out the adjacent walls.

Water coming through the floor or standing at several penetrations, low spots, or corners is a sign of *wet feet*, a code-word for water circling the building and/or rising underneath it. The mechanism could be either a surface or subsurface one, but the likelihood is subsurface since surface penetrations are more likely to occur at one or two places. Also water issuing from around pipes and columns indicates substantial hydrostatic pressure that often arises from a column of water standing some distance above the floor slab. The water could be next to the building or some distance away.

Sometimes distinguishing between clean and dirty water on the floor will help in the diagnosis. Clean water may be flowing quite slowly and coming from an extended distance. Dirty water is more likely to be percolating through the adjacent backfill, which has never stiffened, and flowing fast enough to carry soil particles in suspension.

While such an occurrence is indeed rare, a house floor slab could burst if the underlying pressure exceeded 50 pounds per square foot, and the slab bore no other load such as equipment, partitions, furniture, and the like. This figure comes from estimating a 4-inch-thick slab at one-third the weight of a cubic foot of concrete (150 pounds). If the pressure from underneath increases beyond an unknown practical limit, preexisting gaps at penetrations—if any exist—allow the influx of water, which in turn relieves (lowers) the pressure outside.

Concrete masonry unit walls also may be subject to such implosion, but they usually are saved by three factors:

- The water from one source flows around and under the building as it rises to some height against one wall.
- The foundation walls are surcharged by vertical (floor, wall, and roof) loads that strengthen the resistance of the wall.
- Walls are braced laterally by partitions or the ends of them.

As discussed before, water problems are best solved at their sources. If water continues to flow into a building for several hours after a storm passes over or for several days after a long rainy spell stops, it may be coming from a broken storm drain nearby. Of course a broken sanitary pipe or water service could be the culprit, but they work their woes during dry periods. In the case of the former, the odor is unmistakable. If the utility mechanisms are eliminated, a groundwater source remains.

Treating a low spot in the floor (as a single location) with a sump pump alone was discussed above. If the house already has a full-perimeter dewatering system in place that has succeeded for a few years and the isolated spot is a new symptom, do not install another pump. Instead lay pipe and gravel in a narrow channel under the slab between the new wet spot and the nearest location of the perimeter system. Connect the new pipe to the perimeter system. The new pipe and gravel should act as an interceptor for that portion of the inflowing water that has bypassed the perimeter system (either under or over). This phenomenon is possible although it happens to just a small percentage of the total installations.

If wet feet persist in a home where all the roofing, guttering, skin, and yard corrections have been made, probably only a complete perimeter dewatering system will solve the problem. For an area of up to about 1,200 square feet, one pump should suffice. Much larger areas or those that are cut up might need a second pump. If the house has both a crawl space and a basement install the dewatering system in the area that floods, usually the lower one. In the future, the upper area might need a partial or full piping system that drains into the lower system.

If the basement contains a valuable finished area, you should consider a battery-backup sump pump to protect against loss of pumping action if the electricity goes off. The outlets of sump pumps should exhaust far enough away from the building so that neither surface water, nor subsurface water *recharged* from the surface, will circulate back toward the building. Recharge occurs where water ponds or even flows slowly on the surface to such an extent that it infiltrates and percolates through porous soils before most of it can evaporate or run off.

Window wells usually fill up from uncontrolled water flowing into them from above. A common source is an overflowing gutter from the various mechanisms discussed previously. Correcting the gutter and covering the well with transparent plastic should solve the problem. Occasionally after these corrections, the well continues to fill up from water rising from underneath the well (Figure 4-5). (Perhaps the window is tight enough to keep out the water and strong enough not to break from the pressure against the glass.)

Sometimes the gutter was not fully corrected, or a grading problem went undetected, and the runoff flowed down through the soft back-

*Figure 4-5. Water in Window Well*

Water stands 2-feet deep in a basement window well without penetrating through the wall under the window. A black polyethylene sheet is draped down the wall under the brick from the top of the window sill, but it does nothing to keep out the water because the water enters the window well from underneath and rises in it. An oak leaf floats on the surface of the water.

fill around the well that usually is not more than 3 to 4 feet in the ground. Because water weighs 800 times what air does, the atmosphere (or ambience) inside the well permits the water to rise to balance the hydrostatic pressure.

Solving this problem might take different tacks if the water source cannot be either found or stopped. You could install a drain inlet inside the well several inches below the window sill and daylight it preferably downgrade outside the home. But failing that solution, you could extend a pipe from the drain inlet down to the footing drain that should safely discharge the water away from the house. If a footing drain does not exist and a perimeter dewatering system would be prohibitively expensive, install a sump pump in a pit down at the footing level.

Water may penetrate under basement doors or alongside them when they are buried in an areaway. The simplest mechanism is an areaway drain blocked by leaves or silt that forces the water to pond against the door. While a concrete door sill should be at least 4 inches high to give some protection, water might rise a foot or more before entering. Homeowners tend to cover these drains with variously sized and shaped screens that only delay blockages but do not prevent them. The best solution is to check the drain during major rains and remove any litter that has accumulated. If the drain exits to a sanitary sewer or storm drain, the homeowner should flush the drain with a hose at least once a year.

A second mechanism is water infiltrating and percolating under a stairway or through the areaway walls. In the third mechanism water also may rise from underneath the lower platform. These two mechanisms often work together. Once again you should work to locate the source(s) and divert the water away from the home. Filling holes under the upper platform, regrading the backfill away from the wall, and improving an adjacent patio are just three of the corrective measures you could take. Adding to the height of the areaway is probably more expensive, but perhaps necessary for you to feasibly increase the slopes away from the walls. You may wonder if this change will add to the water and soil pressure against an already underdesigned wall and perhaps cause it to fail. If the wall looks stable, another several inches of soil against it probably will not do any significant harm.

## Plumbing and Heating Systems

Water appearing on the insides of homes may also originate there. Examples are leaks from plumbing and heating pipes and equipment. Plumbing and hydronic heating leaks were mentioned early in this chapter under Diagnostic Procedure. They should have been eliminated from your list of causes and mechanisms long before now.

# Treatments of Condensation and Slab Penetration

Usually condensation appears in small amounts such as water droplets and stains at basement corners, damp smells in unventilated closets, and dampness on bathroom walls hours after use. In cold weather, condensation should disappear from heated spaces although some residual molds may remain. In hot weather you can use aluminum foil or microscope slides to explain the symptoms on walls and floors. When either of these two are temporarily attached to a wall for several hours the accumulation of moisture on them tells where the moisture is originating. If the moisture is on the wall side it is not condensation: it probably is penetration through the wall. The solutions for water penetrating walls were discussed in detail earlier. If the moisture is on the room side, it probably is condensation.

Generally you can cure condensation problems by reducing moisture and increasing ventilation. You should recommend to customers that they take the following steps:

- Use a vent fan in the bathroom and kitchen.
- Take shorter showers.
- Adjust humidifiers to inject smaller amounts of water vapor into the air.
- Check that clothes dryer vents are free of lint and are exhausting outside.
- Avoid hanging damp clothes in closed areas.
- Add ventilation to walk-in closets or keep their doors open with the light on during periods of high relative humidity.
- Upgrade attic ventilation.
- Increase ventilation of foundation crawl spaces or condition those spaces the same as finished rooms. Conditioning crawl spaces requires that they be closed to the outside and opened to the adjacent finished room(s) that are heated and/or cooled.
- Perhaps use a circulating fan or two at the far reaches of the crawl space to move air around in it. If the walls are exposed to the outside, insulate them.

Suppose that hardwood flooring laid over a concrete slab on grade (SOG) of a first floor addition warps and separates. If the wood surface is damp, the moisture is probably from condensation. Moisture also may be migrating from underneath. Treat these mechanisms one at a time.

If the condensation can be stopped or reduced perhaps the flooring can be repaired without replacing it. Be sure the walls and ceiling are not leaking air and that they are adequately insulated, including the fenestration. Test the improvements through changes of seasons. If no new symptoms appear, repair the flooring.

If the flooring condition worsens and no surface dampness is seen, the probable mechanism is penetration from underneath. If the flooring is only slightly damaged, perhaps you can improve the drainage adjacent to the slab enough to prevent water from migrating under the slab. Directly improving the drainage under the slab is desirable if it can be done without access through the flooring to get through the concrete.

If the flooring is heavily damaged, it will need removal and replacement. Removal also gives you access to the slab under which you can install drainage in the subgrade as discussed previously. Wait several weeks or months, depending on the climate, to be sure the slab no longer leaks.

Before replacing the flooring take the following steps to protect it in case the slab leaks in the future. Spread a compatible mastic over the slab and lay a heavy waterproofing membrane in it. Carefully fur over the membrane with wood sleepers using adhesive but do not puncture the membrane. Lay strip flooring directly over the furring or lay sheathing to support parquet. Leave a gap between the flooring and the wall all around the perimeter to allow for any future movement in the flooring. Cover the space with baseboard and/or shoe molding. If you can figure a way to leave air spaces through the molding, do so.

## What If Treatments Are Unsuccessful?

Many builders, remodelers, and developers have encountered at least one unsolvable problem. No matter what treatment, curative method, solution, or preventive measures they tried, nothing fully worked although usually the steps taken improved the situation. One excuse for this dilemma is that you or the customer ran out of money before you tried all the treatments. When you are using your funds, the homeowner probably does not care. But some homeowners who have to foot the bills even decide they can live with some discomfort and annoyance when the "permanent" correction price exceeds the temporary repair price several times. An example comes to mind: after living in a home for 18 years, a client had his first wet basement during a major rainstorm. He paid $500 to have the recreation room rug cleaned. When faced with spending $4,500 to install a dewatering system, the only treatment left after others were found wanting, he decided to take his chances with future storms and penetrations since $4,500 would pay for nine rug cleanings.

Part of the hesitation to expend large sums results from the doubt that a permanent solution is really forever. Contractors' warranties for water problems last 1 or 2 years and consultants do not underwrite their advice. Another part of that hesitation results from doubt that the solution will work. For example, in a few cases even the dewatering systems installed in homes do not intercept all penetrations. Water does what it "wants" to do, so it may bypass a system by going over or under it, instead of through it. (Of course, water obeys known physical laws, but if you are not allowed to pick up the house with a crane so you can look under it, you have to conjecture or infer or conclude, where the water is and where it is coming from.)

After a thorough study of a problem, a professional consultant starts with the least expensive and simplest solution that is appropriate. It may not be the lowest-priced method because that one may already have been tried. For example, many wet basements result from dirty rain gutters and/or blocked downspouts. But if that correction was done just a few days before, it obviously need not be repeated.

If success eludes you, restart the diagnostic procedure laid out near the beginning of this chapter. Various things could have gone wrong the first time:

- You may have missed a symptom or condition.
- The owner may have forgotten or ignored a seemingly tiny but important detail.
- You have been unable to connect just one symptom with any mechanism.
- One of the treatments was incorrectly or inadequately done.
- Perhaps a subsequent treatment vitiated a previous one.
- Conditions have changed.

After all your and others' efforts to solve the problem, enough of the problem remains to sour your relationship with a homeowner. What do you do?

First, discuss the whole matter with him or her in a completely frank manner. Admit that while all problems are theoretically solvable, a few are practically unsolvable without major (expensive) removal and replacement. But quickly add that you are not giving up: volunteer to check back on the progress or retrogress of the situation and request that the owner notify you if a change occurs.

Second, offer to share the price of a knowledgeable and respected consultant to be hired by the homeowner (who should be assured that the consultant's loyalty is to the homeowner). A neutral consultant can provide an independent review of the problem and the treatments. For the consultant to give the best advice he or she will need your full knowledge, even any details you might prefer not to reveal. The goal should be to solve a problem, not to fix the blame.

# 5

# *Examples of Problems and Solutions*

This chapter presents examples of residential water problems that show how to apply the information in the previous chapters. The situations will vary in complexity from simple to difficult but not in that order.

## Roof and Chimney

A medium-pitched, wood-shake roof leaked in four places: inside the chimney, around a skylight, along a valley, and around a small plumbing vent. A sloping, cementitious covering called a wash protects the exposed brickwork around the flue liner or liners on the top of a chimney. This covering on a severely cracked chimney cap was made of mortar instead of concrete. The rest of it had to be chipped off and replaced. The correct skylight for the roof was adequately flashed, but where the plastic bubble fit into the metal coaming (raised frame), the sealant had dried out and developed many holes.

The valley flashing extended up each rake a sufficient distance for the pitch, but the rolled edges along each side of the flashing that are supposed to turn back side-moving runoff were missing. The valley did not leak in its early years because the roofing fit tightly to it. However weathering loosened the shakes, and a space developed between them and the flashing. Replacing the flashing along with some decayed shakes solved the problem.

The plumbing vent flashing was installed backwards when the roof was new so that the upper leaf was over the shingles and the lower leaf was under them. The flashing rarely leaked because it was located high enough on the roof to receive rain and snow but little roof runoff. A new flashing was installed correctly.

## Roof and Yard Runoff

### *Roof Runoff*

After having a dry basement for 18 years, a homeowner discovered a flooded basement two weekends in a row during heavy rains. The clean water came through the joint between the front wall and the concrete floor. The volume of water totaled a few hundred gallons by the time he finished cleaning up. The penetration stopped shortly after the rain ended each time.

This problem is acute instead of chronic. To solve it you would look for a simple triggering mechanism by asking questions after the homeowner finishes relating the facts as he or she remembers them:

- Did the homeowner notice water sheeting down the front of the house at any time?
- What direction was the wind coming from and how strong was it?
- When did he or she last clean the gutters and flush out the downspouts?
- Was water issuing from the downspouts, and where did it go?
- Does rain water or melting snow pond at the foundation planting area?

The best approach is to get the answers to these and other questions that arise during the discussion while you and the homeowner are observing the home on the outside.

In this case the owner reported that water was spilling over the gutters (of a narrowly overhanging, shallow-pitched roof), and the downspouts were flowing almost full. The wind was blowing diagonally across the house and gusting intermittently. The owner saw the outflow from the downspouts running away from the home at both corners as it should have because he recently had cleaned the gutters after the leaves stopped falling.

The soil next to the house had minimal downslope away from the foundation, and he had been backfilling with mulch instead of soil. Thus the mulch was much thicker than the recommended couple of inches. The edges of the planters had no barriers to dam the water against the house, but the water spilling over the gutters was infiltrating the backfill and percolating from there. But why was the water spilling over the gutters?

Seen in profile at the eaves, the butt ends of the shingles properly extended over the gutter no more than a third of its width, but the gutter was attached too low on the fascia (gutter board). The gutter placement was verified by a front view of the roof slope in plane: the gutter did not block the view anywhere. Gentle rains would slowly run off the ends of the shingles and into the gutter, but hard rains probably overshot the gutter and forcefully fell to the ground underneath.

Some readers might protest that in parts of the country subject to heavy snows, placing the gutters high on the fascia exposes them to damage or being torn off by snow sliding down the roof rake. The answer is to use plenty of well-maintained snow-guards. As for ice dams, if the home is properly ventilated and insulated at the eaves, the only other preventive measure is electric heating wires laid where ice dams have occurred. But those wires are susceptible to both electrical and physical failures.

A question still remains about the clean water that flooded the basement. Usually the water that rapidly flows into and out of soil backfill will carry some fines in suspension that at least cloud the penetrating water. Often the water will be dirty and not just turbid. Clean water usually indicates groundwater that has come a distance and/or flowed slowly. In this case the backfill may have had a substantial amount of clay that acted as a binder for the remaining soil.

Another indication that the flood originated nearby is that the water stopped penetrating when the rain stopped. If the subsurface water had been flowing more of a distance, probably it would have kept penetrating hours, or even days, after the rain stopped.

## Blocked Downspouts

This example adds another argument for not burying downspout extension pipes without a good reason. To repeat, the first argument stressed that even with routine maintenance underground pipes get blocked by litter, tree roots, settlement disconnections and the like.

The winter of 1993-94, one of the worst on record along the East Coast, set the stage for the second argument. A homeowner's basement suddenly began flooding after a winter rain turned to ice as it landed on frozen ground. She could not remember when the basement had flooded previously. Outside she could not find any obvious symptoms until she chanced to look at the downspout extension pipe outlets at the street curbing. Neither was outflowing. Her husband then remembered that he had installed those pipes with minimal or no slope from the house to the street. Roof runoff water in the pipes was flowing so slowly that it froze solid and completely blocked the pipes. A few days after the next winter-thaw the pipes began flowing again and the basement stopped flooding.

## Yard Runoff

An old home was being remodeled and just before construction was completed, the basement was flooded during heavy rains. A consultant routinely inspected the usual sources of such water and narrowed the mechanisms to roof and yard runoff.

In addition to enlarging the home on the same lot with narrow sideyards, the home owner had added patios in the rear and a long driveway on the right side of the home. The yard on the left side was nearly flat compared with the house and the street, and it could not be improved. The consultant concluded that several downspouts that disappeared underground connected to too few pipes that daylighted near the street.

Water entered the basement on the floor along various walls and came up through a floor crack in the furnace room at the left side of the basement. Inspection of a hole bored through the slab near the crack in that room revealed only damp soil.

The recommendations that follow solved the problem: (a) add enough underground pipes outside to match the total sizes of downspouts that emptied into them and daylight them separately and (b) install a full perimeter basement dewatering system inside the home with the sump pump located at the left front corner of the furnace room.

The homeowner followed the advice, and nothing was heard from the homeowners for 4½ years. The property was subjected to many rains and snows during that period. At the end of that spring, water penetrated the basement inside the room through the same crack as before. (The crack was never repaired.)

Upon request the consultant met the waterproofing contractor at the residence after the latter had already discussed the situation with the homeowner. The waterproofer and the homeowner (who was an engineer) had decided that no improvements could be made to the roof runoff. The consultant and the waterproofer agreed that the perimeter system was somehow being bypassed both along the left side and the rear of the room. They also agreed that the solution was to add a couple lengths of perforated pipe with gravel under the slab between the existing system and the wet spot on the slab.

The homeowner, citing the price of the dewatering system, asserted that the waterproofer should do the new work for free because the original system was inadequately fulfilling the warranty. The waterproofer replied that his warranty extended only to workmanship and materials but not to a "dry basement." He also reminded the consultant that he had done everything that originally had been specified.

The consultant mediated between the parties and persuaded the homeowner to pay for the extra work because underground conditions appeared to have changed, and the waterproofer reduced the price.

## Excess Yard Runoff

A person purchased an older, single-family home that was sited part of the way up a steeply sloped lot that was many feet higher in the rear than in the front. The ground continued to slope up from the rear of the lot to the next street. The left sideyard also sloped up into a wooded estate. The home inspector warned of gutter and downspout problems and of a hole in the foundation near grade. These deficiencies were corrected after occupancy but not before spring rains penetrated the basement.

After an especially heavy rain the basement leaked in several places:

- at the wall-floor joint in the left rear corner
- across the rear wall just above the slab
- through the floor at the foot of the stairs from the first floor
- onto the garage floor a few inches below the basement floor

This time the water included smelly red, brown, and green slimes (probably including iron oxides) that were hard to remove from the floor. The penetrations continued for several days after the rain stopped.

Three conditions indicated that the water was also flowing under the basement slab: water issuing up around the stair carriages where they penetrated the slab near the center of the basement, bubbling water at several locations on the surface of the slab, and water flowing through the party wall between the garage and the basement.

The owner was upset because he could not use his basement and was in danger of falling when he stepped on the slippery areas. The slime residues remained for many days, but the penetrating water carrying them gradually disappeared when summer weather became unusually dry.

The mechanism for this penetration is the excessive subsurface water flowing from the wooded and improved areas upgrade from the house with the problem. The ground water was *recharged* from surface penetrations an unknown distance away. As the water seeped and flowed underground it leached out the metallic deposits from the subsurface soil. These colored the water (see the appendix).

Solving this problem might start with getting permission from the land-owners above this property to look for ways to divert surface runoffs from both improved and unimproved areas. In such a situation, even if you could find ways to do so, the other owners probably would not want to make changes at their expense that might benefit only a distant neighbor.

The second treatment, a two- or three-wall perimeter dewatering system (pipe, gravel, and sump pump) under the basement floor might be in order except that the viscous mud in slimes blocks sump pumps. Even high-quality pumps suffer reduced longevity from these effects. Cleaning the pit and flushing the pump with fresh water should extend the pump's life, but occasionally replacing a pump is a small price to pay for a usually dry basement. If the basement is especially large or cut-up, you need to consider a second pump diagonally opposite the first one. The two pumps would back up each other if the pipes were joined as one system and laid approximately level. If the basement contains valuable items, a battery-operated backup pump should also be installed above at least one of the primary pumps.

If the second treatment were unsuccessful because the basement dewatering system was bypassed by water seeping or flowing under it, an expensive third treatment is possible. It involves building a deep intercepting ditch across the rear lot line and down to the street along the lot line on one side. This ditch would both intercept the surface and subsurface flows and convey them along lot lines to the street in front of the home.

The ditch would start at the surface, be at least 1-foot wide, and continue to a depth below the basement floor elevation (Figure 3-1). The ditch would contain large gravel all the way to the bottom. The ditch would be too deep for water at the bottom to flow by gravity to the street, and pumping is not a practical alternative. Therefore, to use gravity, the ditch would need perforated pipe laid in the gravel at different heights above the bottom. You determine its vertical placement by *differential leveling*. You would work up from its outlet at the street (using a transit or level, folding ruler, strings, stakes, and so on). The pipe should slope up from the street minimum average of 1 percent or about ⅛ inch per foot throughout the ditch.

Across the rear lot line on the house side of the ditch, a plastic membrane should line the wall to prevent exfiltration through the ditch wall. The membrane theoretically should also force in-flowing water to rise up in the gravel-filled ditch to where it would be intercepted by the pipe. Unless a special condition requires it, do not cover the bottom of the ditch with the membrane because some percolation out the bottom is permissible. If the ditch were dug into a highly permeable material, such as sand or gravel that outletted at the house foundation, the ditch would need a bottom liner.

Because the third treatment probably would cost thousands of dollars, it would be prohibitively expensive for many homeowners. Therefore, if only clear water penetrated and found its way to a floor drain, they might want to take their chances on the infrequency of long or hard rains and deal with the consequences during worst-case conditions.

### *Multiple Problems*

A first time homeowner had lived 2-years in an 8-year old home. After a severe rainstorm he discovered that water had spread inward at multiple corners of his basement. He found that his sump pump had failed and had it replaced. Soon after the replacement, another large rain occurred and his basement reflooded.

A consultant inspected the backyard with a long reach up to the adjoining park land. He discovered that the swale across the backyard near the home was inadequate to divert surface runoff around the house and down the sideyards to the street. The consultant recommended major improvements in the backyard grading.

The previous owner had added a large, intricately designed screened porch, deck, and steps down to grade at the rear of the home. Grading under the deck was flat and exposed to backyard runoff. It also received a corner downspout's outflow because the short shoe pointed backward and had no splashblock under it.

Just before leaving the site the consultant remembered the sump pump at the rear of the basement and began looking for its outlet. The owner remembered a polyvinyl chloride (PVC) pipe that exited the rear foundation wall and pointed down to the backfill. The consultant found this pipe behind the wooden stairs leading down from the deck. He determined that it was the sump pump outlet pipe.

This example reminds you to direct yard and roof runoffs away from a home in all directions. It also teaches you to direct sump pump outflows away from the home and outlet them at some distance so that they do not recirculate between the sump pump and the backyard.

# Skin Leakage

Skin leaks can occur at every detail where materials or surface alignments change. The following examples are taken from different homes to illustrate the many places that leakage occurs:

- Windows were set in framed openings that were too large to be caulked, but somebody tried it anyway. The solution was to remove the caulking, install jamb fillers, and seal the remaining narrow gaps with caulk.
- A two-story brick home had a Colonial-style entrance feature around the front door. Because the owners did not want to see flashing at the head of the feature, a thin bead of caulking filled the gap between the back of the feature and the surface of the brick. Within 6 months after occupancy, the owner found water stains on the hardwood floor of the foyer. She remembered the caulk joint and started checking around the feature. It was still adequate, but she noticed the front door was getting hard to close. A building inspector discovered that the front stoop was rotating away from the home and moving the door and brick veneering with it. Blowing rain was entering the home between the bottom edge of the front door and the metal threshold. A weather stripper installed a water table (Figure 1-1) on the front door as a temporary measure until the structural corrections could be made.

- A custom home with a complicated gable end had rain leaks inside under the triangular louver when the wind blew toward that location. The wooden slats were normally angled down and showed no signs of water passage. But where the louver fit into the large brick gable end, ¼-inch gaps were obvious up close but not from the ground. A high-quality sealant was the answer.

# Paint Peeling from New Home Siding

A new frame house in a rural area was sited on a tableland with no trees close to the building. The house had two different shapes of horizontal poplar siding, bevelled and dropped. The builder reported no problems with the bevelled material even after 2-years, but the dropped siding began buckling, cupping, and splitting 6-months after it was installed. The builder replaced the dropped siding with new, kiln-dried material that he routed across the back to relieve buckling stresses, and back-primed before attachment. After he installed the new siding he finished the front faces with two coats of paint. Much of the new dropped siding cupped and buckled again within several months. The owner who was understandably disturbed hired a consultant to study the problem.

The cross-section of the dropped siding wall from inside to outside was as follows:

- Drywall nailed to 2x6-inch studs
- R-19 insulation batts between studs
- ⅝-inch plywood sheathing covered by number 15 asphaltic roofing felt
- dropped siding tightly nailed over the felt and into the studs
- all siding joints and flush nail heads sealed and painted

The cross-sections of the bevelled siding walls were the same except for a thin air space behind the siding.

A test hole through the dropped siding half-way up one wall revealed a damp vapor retarder (the roofing felt), and soft or decayed sheathing. Obviously moisture from inside the home migrated into and became trapped inside the cross-section from where it attacked the back faces of the siding.

The treatment included—

- removing the dropped siding and the roofing felt
- attaching a breathable sheeting to the plywood sheathing
- furring over the sheeting and into the studs with ⅜-inch thick strips that were thin enough to retain some reveal at window and door openings, but thick enough to allow the passage of air for evaporating the moisture from inside
- attaching the back-primed siding with nonrusting nails
- installing ventilators at the cornice where the top siding board abutted the frieze, and at the bottom board where it rested on the water table at the brick foundation (Full-length screening might move a little more air at the top board, but individual ventilators were much less labor-intensive.)

- priming the new wood and finishing it with two coats of paint (Three coats of good-quality paint should last twice as long as two coats.)
- advising the owner to repaint no oftener than 3 to 5 years to reduce the buildup of paint

# Paved Deck Leakage into Rooms Below

Two large luxury homes had rear decks that were paved with either flagstone or brick and flagstone over structural concrete slabs. One leaked into a garage below and the other leaked into an interior swimming pool area below. The membranes installed between the pavements and the slabs were not continuous and had lost their integrity. The water caught by the membranes could not fully exit through the masonry surrounds at the edges of the decks.

To stop the leakage the builder sealed joints in the flagging and around deck penetrations such as posts and columns. He removed stones and repaired and/or replaced the underlying membrane. These various treatments lasted several months before the leakage began again.

The consultant observed the symptoms after at least two attempts at solutions and discovered that one or two components were missing: (a) the horizontal drainage materials, such as geocomposite boards, that should have been laid between the bottom of the pavements and the top surface of the membranes and/or (b) the exit weeps from the geocomposites through brick borders or stone caps at the edges of the decks. The drainage materials were obviously missing when the leakage resumed downward. The weeps also obviously were missing when the downward leakage stopped, but the vertical surfaces of border walls were stained, had soft mortar, or were covered with leached residues from the cementitious joints. The weeps also needed through-wall flashings with drip edges that extended beyond the faces of walls to eliminate the water trails on the walls.

# Excess Dewatering

This example shows that too much dewatering of a site can increase structural damage rather than alleviate it. This home was more than 20 years old and situated in an area of *expansive clay* (see Soils in the appendix). This clay is sensitive to moisture: when saturated the soil expands in volume and applies strong pressure to overlying or adjacent structures. A heavy structure, such as a tall office building can resist movement, but expansive soil can move portions of a home in different directions at variable rates.

In contrast, when the moisture naturally leaves the soil, the remaining soil shrinks and removes support from the structure it had formerly forced to move. Such cycles over time cause inflexible materials to crack, separate, and partially reclose. The results can vary from treatable cracks to structural failures.

In this case cracks in foundation walls and at the basement fireplace had become annoying but not threatening. These vertical and stepped cracks measured up to ¼ inch in width. The sizes of the cracks responded to gross weather changes: during the late summer when local weather was droughty, the cracks usually opened their widest; during the late winter after rains and snows, the cracks were at their narrowest.

Following the usual recommendations of a home inspector and a soil scientist, the homeowner had taken steps to (a) discharge roof runoff well away from the foundation corners and (b) prevent the accumulation of water adjacent to foundation walls. He extended downspout shoes 10-feet or more and increased the slope of foundation planting areas.

After the next dry period, the homeowner was disappointed to find most of the cracks wider than ever before. He showed the third consultant what he claimed he had been advised to do. He also pointed out the areaway to his basement where he had not changed runoff conditions because he did not know what could be done. Significantly, the wall cracks nearest the areaway had not widened as much as the others. The consultant concluded that too much of the roof and yard runoff was directed away from the home, thus erring on the dry side of the optimum moisture content for this particular soil combination and increasing its shrinkage.

The new advice was to shorten the downspout shoe extension pipes to allow some of the runoff to stay near the building, but to still direct most of the runoff away from the foundation corners. Slopes of the adjacent planting areas were to be left as he had modified them. The areaway would remain as it was.

Many months later the homeowner reported some improvement in the cyclical changes of the cracks, to the extent that he stopped considering structural repairs and decided to "live with the situation."

# Recharged Groundwater

After a severe winter the basement of a 30-year old home that had been dry for 18 years began shipping water. Relatively clean water inflowed at both rear and front walls of the recreation room down at the floor line. The inside surfaces of finished walls were dry.

Elsewhere in the basement inside a finished bathroom, the bathtub waste pipe penetrated an open hole in the floor slab. Clean water stood in the hole at a level just below the bottom of the slab. The owner had not seen any water flow up and out of the hole, but on the advice of the consultant she tried evacuating the water in the hole with a small sump pump. For periods of 30-minutes or so the pumping created drawdowns, but when the pumping ceased the water level soon rose to its previous height and stayed there. This test proved that water was inflowing under the basement floor from an unknown source.

The basement exited at the rear into a deep areaway. At the first riser above the lower platform, clean water issued from a hole in the outboard wall (paralleling the steps). It flowed into a nearby areaway drain inlet.

A swimming pool behind the home was not leaking. Just beyond the pool a 5-foot high retaining wall bordered a long backyard that sloped up into neighboring yards and to homes fronting the next street up. The lay of the land indicated that abnormal precipitation the previous winter had recharged the long reach behind her house to such an extent that subsurface water flowed toward her home. Those flows at first probably traveled under the home until they were dammed by the front foundation wall acting as a cut-off. As the water accumulated and rose under the home it ultimately penetrated at various locations around the basement.

Little could be done outside to prevent recurrences so the only remaining treatment was to install a sump pump in a pit under the floor to maintain a drawdown in preparation for any future inflows. Because the problem was an infrequent one, a drainage pipe under the floor along the perimeter walls was not needed at this time. But if the sump pump could not keep the whole basement dry, such a pipe would become necessary at certain locations.

## Groundwater Penetration of New Addition

A homeowner contracted for an addition to an existing 25-year old bilevel home on a steeply-sloped lot that was uphill from a creek that flowed on the opposite side of the street. For many years water flowed under the home in gravel trenches that outletted in the front yard.

The addition was across the rear of the right half of the home, and was built over a sloping crawl space that was shallow at the rear and deep at the front. At the front of the addition a new circular stairway gave access to the existing lower level that was extended enough to include the stairway.

The grading of the rear yard was not changed, so the toe of the steep slope came within 3-feet of the rear of the addition. To protect the foundation of the new addition from surface and shallow subsurface runoffs the contractor installed a trench drain around all three sides of the new addition at the base of the foundation. This drain ultimately daylighted at the street where it proved to be working.

After a home had endured the wettest March in over 100 years, water penetrated at the base of the masonry pit containing the stairway and flowed across the lower level slab damaging the new parquet floor and subsequently flowing under an auxiliary exit door.

Even though the new foundation drain was working, the consultant believed deeper flows of water in the hillside passed under the drain and attacked the CMU walls of the stairway pit. The consultant wanted the contractor to improve the dampproofing on the positive sides of the stairway pit walls, but the latter claimed there was too much competent rock to remove and too little space to work over it.

At the consultant's urging the contractor installed a sump pump under the floor of the stairway pit, but he did not also install a trench drain under the floor that would connect to the sump pit. As a result, future penetrations through the crawl space were only partially intercepted by the pit and pumped out. The rest of the penetrations again flowed out onto the floor. The owners hired another contractor to install a trench drain under the floor to intercept all the penetrations and lead them to the sump pit.

## Tidal River Penetrations into Basement

Along the shore of a large East Coast river, tidal currents overflowed the bank and pushed into the basement of a home. At first the owner considered installing a bulkhead along the shoreline to ward-off the water, but since other neighbors did not have her problem without the bulkhead but would have it with a new bulkhead protecting only her property, she looked for another solution.

The waterproofing contractor she hired installed expansion tanks (similar in purpose to those found in hydronic heating systems) underground in a line paralleling her front foundation wall and some 20 feet from it. The contractor designed the size and number of tanks to contain the estimated volume of water in a tidal surge. The tanks had covers with holes large enough to capture water flowing over them. The tanks were flush with the surface but were not connected to each other. The walls of the tanks had holes that acted as reliefs (weeps) to the surrounding soil. Ejector pumps could have been installed if the tanks ever overflowed, but the pumps were never needed.

# Malfunctioning Basement Dewatering System

A several-year-old, interior, two-story townhouse began shipping turbid water into the basement. The sump pump ejected the water that flowed out of the pipes into the sump pit, but water still flowed out onto the floor from a side wall at the rear of the basement. The original owner checked that roof and yard runoff were not apparent sources of the penetrating water.

This energetic young man decided to excavate across the rear of his home several feet down to the footing. He left the excavation open until the next major rain when he discovered water standing on the footing. He could not detect where water was entering the excavation, but he believed it was rising from under the home.

At this point a consultant saw the situation and listened to the owner recount his lack of progress. The indications were that the dewatering system was only partially evacuating water from under and/or around the home. Following the consultant's advice, the owner pushed his garden hose alternately into the pipe exits at the sump pit as far as he could, and alternately turned on the water. During both tests, water began issuing from the slab-floor joint at the same place on the side wall that water entered the basement during a rain. The obvious conclusion was a blocked perforated pipe under the floor in that vicinity.

The homeowner broke through the floor and discovered a crushed pipe section that he replaced. Before replacing the floor section he waited for tests of the repaired system by minor and major rains. No water flowed onto the floor, but the major rain substantially increased the flow out of the pipe end that was closer to the repair. The consultant then agreed with the homeowner that the backfill and the floor could be replaced.

# Plumbing, Heating, and Cooling

Various conditions will cause water supply pipes under pressure to leak into their surroundings. At atmospheric pressure (*ambience*), even drain pipes will leak under the conditions listed below:

- They are penetrated by a nail, and it is removed.
- They split because the contents freezes, and then the contents thaws.

- Fittings or the pipes themselves loosen or break.
- They corrode or wear too thin.

The first three causes usually are obvious, but the fourth sometimes is more obscure. Finding the exact location inside a wall, ceiling, or floor may not be easy. But the mechanism should be evident from the continuing leakage that is not correlated with weather changes.

## Plumbing

A split-level home had a wet crawl space under the living room wing. Most areas of the crawl space were accessible or observable. A wooden partition had decayed at the bottom plate and partway up the studs, water stood alongside or on top of an interior wall footing, and the foundation concrete-masonry-unit (CMU) wall under the kitchen was heavily stained and coated. A waste pipe under the kitchen floor was leaking profusely (Figure 5-1). Streams of waste water from the leak stained the wall and coated it with brown grease globules.

From the appearance of the wall, the pipe obviously had been leaking long enough to spread water to several locations in the crawl space. The solution required several repairs and cleaning.

*Figure 5-1. Leaking Waste Pipe*

*In the crawl space of a home the plastic waste pipe from the sink above is leaking a lot. Note the grease globules sticking to the CMU foundation wall and hanging from the pipe joint near the rim joist.*

## Heating

A 30 year-old slab home on the peninsula south of San Francisco was heated by a radiant hot water system. During the heating season, the floor between area rugs became stained, apparently from water. The home had no previous problems with leakage from underneath, and no symptoms of roof or skin leakage could be found. Test holes drilled into the slab revealed deteriorated copper supply pipes that had corroded from the chemical reaction between the pipes and the enclosing concrete. The owner abandoned the copper pipes and had plastic supplies laid in new channels routed into the slab.

## *Cooling*

The owner of an older home complained of a wet basement wall only on one side of the house during the summer. It never fully dried out through the winter, but it was worse in the summer. The owner was unsure whether the floor also got wet during heavy rains. The home showed signs of neglected maintenance, but it had no obvious symptoms of roof leaks or yard drainage problems.

Outside the wet wall at the sideyard, undergrowth grew around and over a window well. Inside the undergrowth the condensate drain pipe for the central air-conditioning symptom emptied directly into the window well. While this drain may only have issued a teacup of water every hour or so, over several years the accumulation filled many cells of a CMU wall. After the condensate pipe was moved to its proper location, the wall took many months to dry.

# Condensation

A several-year-old townhouse was so humid inside during the winter that the owners became concerned. When the consultant arrived, he could see through his car windows that ice had formed over the front door and trailed down the front of the two-story home. During the initial discussion, the owners mentioned that their asthmatic child required a high relative humidity so they had set the furnace humidifier to its peak delivery rate.

Feeling all the moisture in the air and seeing it on windows, the consultant searched for damage. In the attic under the insulation he found water running between the bottom truss chords. Although the humidifier was in the basement plenty of water vapor had migrated through two levels. The combination of ventilation and insulation in the attic cooled that vapor enough to condense but not to freeze it. Luckily the drywall was not yet soaked through.

At the end of his inspection the consultant strongly recommended that the owners take the following steps:

- Lower the setting of the humidifier immediately.
- Mop the upper side of the second floor ceiling.
- Open all the windows during the next winter thaw.
- Again and again open all the windows as often as possible during windy, warm days the following spring because nature works better and faster than dehumidifiers.

The examples in this chapter provide the solutions to a broad spectrum of residential water problems that have occurred to a variety of houses. They do not cover every problem or situation that may occur to a home. Instead they serve as guides for you to analyze and solve any water problems you may encounter.

# *Appendix*

# Soils, Hydraulics, and Hydrology

This appendix will give you some of the theory behind the practical treatments for the problems this book already has examined and for problems you might encounter in the future.

Why include theory at all? Why not just write the solutions for all the problems seen or imagined? Attorneys, doctors, engineers, and other professionals will answer that theory is the basis for their practices. Most of the time they can solve problems and treat patients using their everyday knowledge and experience. But once in awhile they are stumped by a different situation. Then they must look at a book, do some hard thinking, and perhaps consult a more experienced practitioner. Knowing theory (a) alerts them to the variations in a situation and (b) warns them when a standard solution is not likely to work. Applying theory helps them to devise a solution that will work.

## Soils

Soil is a little like the air people breathe: most of the time they pay scant attention to it, but when something goes wrong, they become attentive. Poor soils have the potential for damaging structures built on them. Examining the technical aspects of soils will help readers to better understand specific problem soils. Discussions of the practical applications of this theory will follow later.

Soils interest agronomists, geologists, and engineers for different reasons. Agronomists (soil scientists) classify soils according to the relative percentages of minerals, organics, water, and air. They are primarily interested in soils for agricultural purposes. Geologists see soils as the residues of rock changes and study their development into forms or topographic features (morphology).

Engineers and engineering geologists—and by implication builders and remodelers—are primarily interested in soils' capacity to carry structural loads. Loads are transferred from buildings to the earth underneath them via some type of foundation. Most construction people know that taking a chance on the strength of a human-made fill, even after many years of in-place (*in-situ*) natural settlement (consolidation) could result in severe differential settlement of a home. (If a building settles as a whole, the problem is more cosmetic than structural. However, such settlement could require structural repairs that would cost much more than the site improvements before construction.

Certain natural soils that may have been in place for millennia also are not fully adequate to permanently carry the relatively light loading of a home (see Chapter 2 for examples of these problem soils and conditions).

The first section of this appendix will look at some engineering and other information about soils that builders and remodelers use. As you can see later, these concepts overlap:

• how they developed and are continuing to change
• how they are classified
• how water affects them
• how they are described by physical and chemical properties

## Origins of Soils

The earth's surface or crust is composed of soil, rock, or rock turning to soil. Water lies over about 70 percent of the earth's surface. The earliest soils came from rock weathered in place by climatic changes (precipitation, wind, temperature) and geologic activity. Since then, soils have been modified by decayed organic matter, organisms, time, slope, chemicals, other minerals, and more geologic actions such as seismic and volcanic activity. These processes continue for both soils in-place and transported soils.

## Classifications of Soils

For precise engineering purposes and based on laboratory tests, soils are classified according to their grain sizes and their behavior based on water content.

## Grain Sizes

Many soil-sample tests include a particle-size analysis (gradation curve or mechanical analysis), an example of which is shown in Figure A-1. However sediments, soils, rocks, and minerals are all different and have different classification systems. Geologists, agronomists, and engineers have slightly different measurement systems for particle sizes.

Going from the largest to the smallest, rocks and soils are named in Figure A-1 as follows:

• Boulders are larger than 12 inches.
• Cobbles are from 3 inches to 12 inches.
• Coarse gravels are from ¾ inch to 3 inches.
• Fine gravels are smaller than ¾ inch but larger than a No. 4 (4.75 mm) sieve.
• Coarse sand is smaller than a No. 4 (4.75 mm) sieve but larger than a No. 10 (2 mm) sieve.
• Medium sand is smaller than a No. 10 sieve and larger than a No. 40 (425 microns) sieve. (A micron is 1 micrometer or one-millionth of a meter.)
• Fine sand is smaller than a No. 40 (425 microns) sieve.

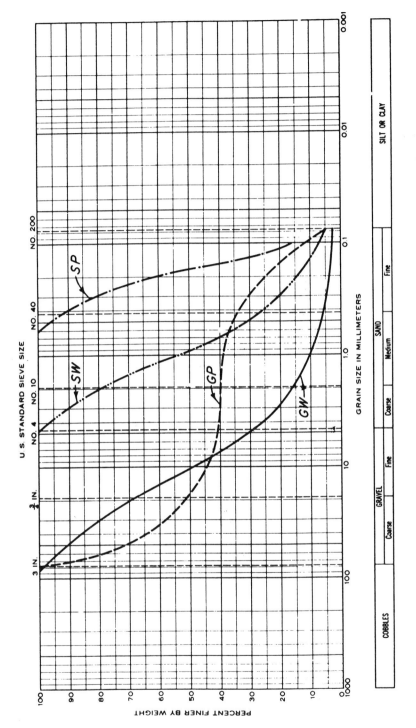

*Figure A-1. Curves of Typical Well-Graded and Poorly Graded Soils After a Particle-Size Analysis*

*Across the top of the graph, screen sizes are in inches, and sieve sizes are in standard numbers. Across the bottom opposite the various screens and sieves, the respective sizes are in millimeters. The bottom scale is logarithmic, which explains why the numbers are closer together as they get larger. The left border indicates the percent passing through the various opening sizes; the right border indicates the percent retained on the various opening sizes. Note that the percentages increase upward on the left so that the curve is cumulative.*

Source: Soils Engineering. Sect. I, Vol.I, (Fort Belvoir, Va.: U.S. Army Engineer School, 1971). p. 37, fig. II-1.

- Silt and clay are both smaller than a No. 200 sieve, but differ according to other criteria.

The following discussion about how soils get their strength will help you to identify and compact materials that do not yet have enough strength to carry the loads of the houses you propose to erect on them.

Granular materials are considered cohesionless (without cohesion). They gain strength when grouped together by natural consolidation or by mechanical compaction. Sharp particles fill the gaps between other particles, squeeze out the air, and increase the density (weight). Ideally granular materials do not have a chemical or physical attraction for each other.

## Shear Strength

Granular materials have a resistance to deformation from loads imposed by external forces. For example, if a small, round rock breaks when it is caught between a steel roller and a concrete pavement, it is likely to break along a diagonal line. In the study of mechanics of materials, this failure is called shear because it is a combination of compression and tension caused by the rock being rolled as it is crushed. The shear strengths of granular materials vary according to their water content: sands, for example, generally become weaker as they become wetter.

## Cohesive Strength

In contrast to a granular material, a clay's strength derives from *cohesion*. This intrinsic strength results from the clay particles' physical and chemical attractions (binder) for each other regardless of their moisture content.

## Plasticity

A material's moldability when dry, moist, or wet is its *plasticity*. It is the other part of the classification scheme for fine-grained soils. Think of a soil sample as a piece of putty starting out dry, to which you continue to add water as you remold it.

A laboratory test for plasticity is based on the Atterberg limits (established states to determine a soil's *consistency*, its resistance to deformation). These states range from solid to semi-solid, through plastic, and on to liquid. They are bounded by the shrinkage limit (SL) at the dry extreme and the liquid limit (LL) at the wet end. The shrinkage limit is the water content of a soil sample at its minimum volume (when more drying will not decrease its volume). The liquid limit is the maximum water content of a sample when it is about to change from a plastic state, such as peanut butter, to a slurry such as soup. Between the shrinkage and liquid limits is the plastic limit (PL).

The previous discussion on plasticity is intended as a brief orientation. It will come in handy if you discover a problem soil on one of your jobs that requires testing. When the results come back you will have an idea of what they mean and thus be in a better position to decide future actions.

### *Physical and Chemical Properties*

Soils and rocks have many such descriptive terms, but only a few apply to construction: color, consistency, structure, drainage, and slope. This appendix will address only the first three here because the last two were thoroughly discussed in previous chapters. As noted before, any one of these properties could be your first clue to potential problems.

Builders and remodelers can learn from the color of a soil something about its present condition and past history, especially the effect water has had on it. For example, on the surface a dark-colored soil, say brown or black, usually indicates a high organic content. But during an excavation if you uncover a dark-gray material that is not mucky and does not smell, you should suspect repeated water penetrations, perhaps by a rising water table. The dark-gray color is the symptom of an oxygen shortage caused by the water penetrations; a brighter color usually indicates a normal oxygen content. (In older books you might find this dark substance called a gley soil.)

The term *consistency* refers to a soil's resistance to deformation or rupture. It is usually used to describe fine-grained soils (silt or clay). It speaks of the cohesion that the soil particles have for each other. Soils with extremely small pores have the greatest cohesive forces. Many common and technical terms are used to describe a soil's consistency. According to American Society for Testing and Materials, terms used for describing in-place consistency are soft, stiff, firm, and hard. Other terms such as brittle, crumbly, tight, sticky, friable, and plastic describe related properties but do not have specific definitions, which the consistency terms do. Good bearing soils are more likely to be stiff, hard, and firm.

*Structure* describes how individual particles aggregate or compound into larger masses. The four primary types are platy, prismatic, blocky, and spheroidal (Figure A-2). For structural considerations only the first and last are of major concern: the first because it is the structure of clays; the last because it is the structure of granular material. The platy structure of clay plays a major role in its water-holding capacity that leads to problems. Sharp-edged granular material that stays dry usually makes a good foundation subgrade because its individual particles fit tightly together.

### *Pressure*

If backfill becomes saturated, the soil's total lateral pressure could double or even triple. If it does, a foundation or retaining wall could break.

# Hydraulics

The engineering application of hydrology, fluid mechanics, and other sciences is called *hydraulics*. (Hydrology is discussed later.) It deals with liquids at rest and in motion and with their resultant pressures. In these pages the discussion of hydraulics concentrates on drainage, including water flows over land and in channels and pipes.

Modern living requires systems to collect and convey water both on the surface and underground. Before humans started felling trees, changing the land's topography, and erecting buildings, nature had worked out the dis-

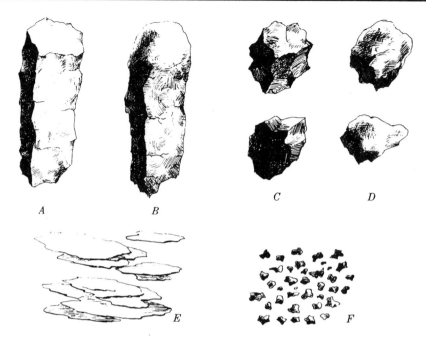

**Figure A-2. Examples of Soil Structures**

*Some of the types of soil structure include (a) prismatic, (b) columnar, (c) angular blocky, (d) subangular blocky, (e) platy, and (f) granular.*

Source: *Soil Survey Manual*, Agriculture Handbook 18 (Washington, D.C.: Soil Conservation U.S. Department of Agriculture, 1951), p. 227, fig. 44.

posal of precipitation. Generally the portions that did not enter the earth flowed overland to various bodies such as streams and creeks, rivers, ponds, lakes, bays, and oceans. Although that disposal system still works, it does not work nearly as well where people have greatly increased the volumes and speeds (velocity) of flows by "improving" the land for agricultural and urban developments. When the water flows faster, it does more damage to land that is denuded of trees and natural ground cover. While the faster water flows over paved and roofed surfaces may not damage them as extensively as the flow over unpaved land, the flows that have become concentrated on pavements can do much harm to the ultimate outfalls that receive those flows.

## *Fluid Mechanics*

### Pipes

When a pipe is full, the flow of water in it is faster near the center because the wall creates frictional resistance (*viscosity*) for the water adjacent to it. For practical purposes but contrary to strict hydraulic theory, the pipe should be so large that it rarely flows full for two reasons:

**Blockage**—Pipes often become at least partially blocked from sediment deposits building up within the pipe and from debris, animal nests, vandalism, and impact damage. While natural flows might flush them out, they are unlikely to do so. Pipes also become misaligned from structural changes caused by different rates of settlement. Thus in major storms, pipes can overflow and wash-out the backfill around them, or they back-up into inappropriate places.

**Design**—The frequency of a major-storm return period and the intensity of a specific storm strongly influence the design of storm drains. Some jurisdictions allow storm drains to be designed to handle a 10-year storm. Statistically but not necessarily actually, a storm of this magnitude could occur on the average once every 10 years. Thus, a 10-year storm has a 10 percent chance of occurring in any particular year.

So what happens when a 50-year storm arrives much earlier or more frequently than expected? If a storm drain pipe is located under a street, the overflow might cover the street and perhaps flow over the curbing to inundate the adjacent yards. If the pipe is in a critical location, such as alongside a home, a flooded basement might not be the worst result: a foundation might fail.

Full flow in a pipe at low velocity usually is streamlined because the individual elements or particles flow straight (*laminar*) without bumping. Some factors that lead to turbulence (seen as swirls and eddies in rivers) in pipes are velocity increases, changes in shape, roughness, obstructions (such as deposits built up within the pipe), and curvature variations. *Viscosity* is the resistance of a liquid to flow.

## *Other Characteristics of Water*

Discharge is the term used in measuring a given volume of water passing a certain point in a specified time period. Examples are gallons per minute (gpm) and cubic feet per second (cfs). Continuity is the theoretical principle that requires that inflow and outflow from a pipe or channel be equal. But to maintain continuity when velocities naturally vary, the cross-sectional areas also vary. The relationship between them is inverse: if the cross-sectional area decreases, the velocity of the water moving through the pipe or channel must increase and vice versa.

The pressure of flowing water (called *hydrodynamic*) is lower than the pressure of the same water standing against a barrier or standing full in a pipe (called *hydrostatic*). When water that has been standing in a container or a pipe begins to flow its internal pressure drops. Thus a weakened water pipe or hose that is nearing failure ruptures when the valve controlling it is suddenly shut. (An automotive example is the weak hose carrying cooling fluid in a car that fails when the running engine is turned off.)

Underground sources of water, whether natural or human-made, usually create only static pressure (discussed below under Hydrology) against walls or slabs. The force exerted is a function of the altitude or depth of the standing water. But if the water is fast-moving, such as from a flood or a broken water main, the pressure against walls or slabs includes a dynamic component. This substantially increased force often causes major damage because force increases to the square of a fluid's velocity. For example, if the velocity doubles, the force quadruples.

Water hammer can occur when a valve is closed too quickly on a pipe containing flowing water. Because of the deceleration and the sudden stop of the water, a wave action develops that travels to the end the water was flowing toward. When the wave stops in that direction it rebounds and flows toward the other end, from where it rebounds again, if it reaches that end. The pipe reacts by vibrating or surging against its restraints and making an unmistakable racket. Under severe conditions a pipe may separate from its clamps, fracture at connections or joints, or even burst. Installing air-filled expansion chambers at some of the fixtures of new work can prevent water hammer.

Freezing water increases its volume by 10 or 11 percent. Thus the phenomenon of boat docks rising higher above the surface of freshwater lakes during the winter. More serious for your purposes is the blocked downspout that fills up with rain or melting snow and freezes. The obvious symptom is the bulge that looks like a snake that has swallowed a small animal. Another result that is harder to see is the sprung seam at the back face of a downspout. Gutters may tolerate frozen precipitation, such as occurs during ice dams (discussed at Chapter 4) because they are open channels.

Of course, you will use adequate insulation around water supply pipes in exterior walls in cold climates to prevent them from freezing, bursting, and leaking.

## Erosion, Sedimentation, and Deposition

Runoff erodes when it flows too fast. The velocity varies with the particular conditions of inflow, channel geometry (discussed below), and bottom material. Eroding water carries off soil particles that have been detached from the surface by the impact and splash of raindrops. This water also cuts rills and gulleys into the land (Figure 1-3), from which the residues (sediments) are washed and carried away in suspension. These sediments may be deposited nearby (perhaps just a few feet away) or many miles away. Those deposits end up in bodies of water, too, where they increase *turbidity*, kill fish and other aquatic creatures, and destroy vegetation. Along stream banks eroding water may scour the sides and bottom enough to weaken and undermine the channel, and even relocate it.

## Geometry of the Channel

The sizes, shapes, and conditions of channels affect flow conditions through them. Although a *V*-shaped ditch is easiest to dig with a pick and shovel it generates the fastest flow through it. Because a relatively deep channel becomes narrowest at the bottom, the water is so restricted that it must flow faster to "keep up" with the layers above it. (This effect is in accordance with the continuity principle already discussed under Other Characteristics of Water.) A better shape for controlling the velocity and volume of water is a short, wide *U* (parabolic channel). It works better than a *V* because the flow is wider, more shallow, and thus slower. (Note that a half-circle-shaped *U* is half of a round pipe.)

An efficient channel shape is a trapezoid with a wide, flat bottom and gently sloped sides. You might not use it if you were to choose because of the added expense of labor and materials, but you could compromise on a *U* with sides sloped at a maximum of 2 horizonal feet to 1 vertical foot (a ratio of 2:1).

How rough should a channel bottom be? The rougher it is the more tortuous the water's path, and the slower the flow. Thus up to a point the addition of cobbles or even boulders helps to prevent erosion. The flow reaches that point when the resistance makes the water flow too slowly, pond, and create stagnant pools that might breed mosquitoes. In addition, you should not throw logs in a channel because they could whipsaw flows so much that they scour the sides, create *bars* (deposits) and ultimately undermine the sides.

Channels (including swales, trenches, and ditches) have advantages over underground pipes. One is ready overflow from many locations during the worst conditions. Other advantages are that they are easy to inspect, and you can detect developing problems before they become serious. The cosmetics of some channels create a disadvantage, but a shallow, wide grassy swale can be used in a lawn.

Even when a pipe is well-placed underground, if you envelop the pipe with gravel up to the surface, the gravel will provide excess capacity to safely carry off volumes too large for the pipe. For this technique to work you need to take certain steps during installation to release overflows out of the pipe.

You can protect the environs of underground pipes from the consequences of overflows by installing a vertical cleanout that also serves as a chimney drain. Locate it where overflows will not damage adjacent areas and improvements. On residential lots a cleanout every 50 feet is not too frequent. When a pipe gets blocked, a cleanout enables you to clear the stoppage from the middle as well as the ends. Flushing is preferable to snaking, especially with plastic pipes. Advise customers to flush regularly instead of waiting for a blockage. Occasionally, a stoppage cannot be removed and the pipe must be abandoned.

If you are sodding a roadside (right-of-way) ditch that is steep enough to cause fast runoff, the fast-flowing water might strip off pinned sod or undermine wired sod. On top of the sod, you can install velocity checks (temporary check dams) to retard flows. Make these out of 1-foot tall scrap plywood pieces or dimension lumber, extend them across the ditch, and nail them to furring strips or 2x4s driven into the bottom. (A substitute for velocity checks is silt fencing used together with straw bales.)

Between the velocity checks, stake down straw bales that will trap waterborne sediment and also help to slow the runoff. Be sure the baling strings around the bale do not touch the ground (to prevent them from decaying quickly). Of course, the velocity checks can be removed when the sod is growing well. By that time the straw bales already may have rotted.

A controversy sometimes arises over the amount of slope (grade) in both pipes and channels. As noted earlier shallow slopes reduce velocities and in turn decrease the effects of erosion, but grades that are too shallow slow velocities so much that ponding and sediment deposits occur. These two results can lead to build-ups in the bottoms that will reduce capacity. The minimum grade for drainage pipes is 1 percent, and for unpaved open channels it is 2 percent.

### Sizing Pipes

Sizing and locating gutters and downspouts probably will be obvious for most residential roofs in your area. But for complicated roofs get advice from local or national plumbing codes, mechanical design manuals, or other books and magazines. The formulas in those publications deal with such elements as—

- roof area to be drained
- roof pitch (slope)
- rainfall intensity (inches per hour at worst storms)
- thickness and texture of the roof covering

You need to match the areas of downspouts, especially multiple ones, to the area of the outlet pipes serving them. During most low-intensity rains, three 2x3-inch downspouts (totalling 18 square inches) might be adequately drained by one 4-inch round pipe (12.6 square inches). However a high-intensity rainfall probably will overload the pipe. For future safety use two 4-inch pipes (25 square inches) or one 6-inch pipe (28 square inches).

Sizing the pipes for collecting and conveying runoff from more than one lot may be more complex than you will want to attempt. If civil engineers design the systems, they become responsible for the designs. If you decide to do it on your own, consult your state's or a national stormwater management systems design and construction manual (see the selected bibliography). The formulas may vary but most of the manuals will include the following elements:

- watershed (drainage) area
- land use (whether buildings, agricultural, forest, recreation, and so on)
- soil type
- land condition (whether grassy, paved, and the like)
- intensity (rate) of rainfall

# Hydrology

The science that studies water—its sources, locations, and characteristics—is called *hydrology*. The primary source of water on the earth and underground is precipitation. A secondary source underground is *connate* water that was trapped during geologic action, often inside rocks. It may be salt or fresh.

Water above the surface of the earth in the atmosphere exists in three states (phases), often mixed together: gas (vapor), liquid, and solid. Clouds of many shapes and sizes, and at different heights, are just one manifestation of these states caused by condensation.

Precipitation occurs when condensed vapor is further cooled until it forms liquids and solids. Common forms are rain, snow, sleet, hail, and mist. The precipitation that falls to earth acts in different ways: it evaporates and *transpires* (see the following paragraphs), runs off to human-made and natural drainages, accumulates on the surface, and penetrates the soil to shallow and deeper depths. (The sequence of actions is not necessarily in

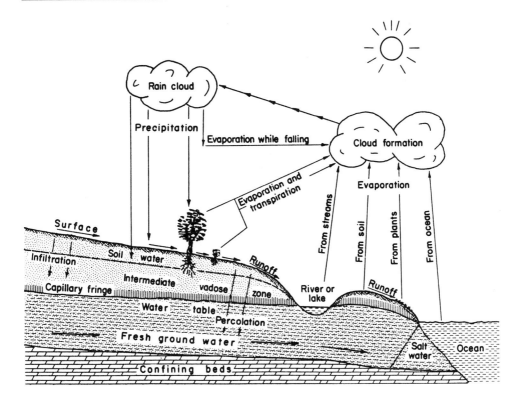

*Figure A-3. The Hydrologic Cycle*

Source: *Ground Water and Wells*, ed. by G. F. Briggs and A. G. Fiedler (St. Paul, Minn.: Wheelabrator Engineered Systems—Johnson Screens, Inc., 1975), p. 16, fig. 8.

the order just listed.) The hydrologic cycle in Figure A-3 is a stylized illustration of this activity in more detail.

## Surface Water

Some of the moisture that lands on natural and human-made objects evaporates into the atmosphere depending on weather conditions such as temperature, wind, and sunlight. *Transpiration* is the moisture that has been taken up by plants and trees, but not used by them, and sent back to the atmosphere via leaves. Evaporation and transpiration together are called *evapotranspiration*. Moisture that does not evapotranspire may penetrate the soil surface (*infiltration*) depending on the soil's readiness to absorb water. The water that neither evapotranspires nor infiltrates may continue flowing overland in drainageways to where it ponds or joins other bodies of water on the surface, depending on the topography (*relief*).

## Subsurface Water

Most of the water that exists underground (also known as subsurface water) results from precipitation that has landed on the earth and infiltrated the surface. This water seeps and flows (percolation) underground in vertical, horizontal, and mixed directions. Figure A-4 also shows underground locations and movements of water.

Subsurface water may resemble surface bodies and flows, but you should not refer to it as lakes and rivers. Subsurface water occurs in zones that are roughly divided into subzones that use soil or rock as their conduits. Note that water, air, and soil or rock co-exist. When voids (*interstices*) contain only air or air and some water, the zone is called unsaturated (also called *vadose* by some authorities and *zone of aeration* by others). In the saturated (or *phreatic*) zone the air in the void spaces has been replaced by water.

The unsaturated zone generally contains three subzones: soil water, intermediate, and capillary fringe. The soil water is closest to the surface and is held by *molecular attraction* and *capillarity* for use by vegetation. The soil water subzone most concerns builders and remodelers (Figure A-4).

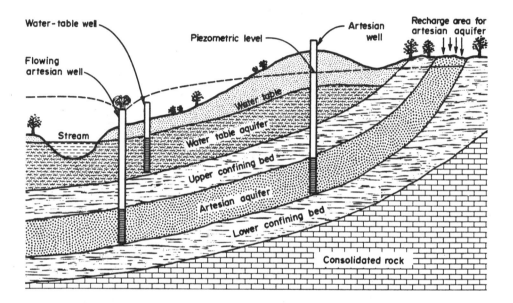

**Figure A-4. *Underground Movements and Locations of Water***

Source: *Ground Water and Wells*, ed. by G. F. Briggs and A. G. Fiedler (St. Paul, Minn.: Wheelabrator Engineered Systems—Johnson Screens, Inc., 1975), p. 17, fig. 9.

## Water Tables

Water in the saturated zone forms layers that may be level or sloped. A water table is the top surface of a particular layer of water. Three different tables can occur (Figure A-5):

- An apparent or normal table stands alone and is at atmospheric pressure when it is free to move in any direction.
- A *perched* table stands above a normal table and is separated from it by an *impervious* layer of soil or rock. One example of such a layer is soil

containing a large portion of clay. Others are *fragipans* and *hardpans* (see Soils). A perched table may be isolated to a small area, such as a house lot.

- An artesian table may exist underground or issue from the earth's surface at flat land or on a slope. Underground a *confining layer* or curtain that is *impermeable* (impervious) enough to prevent leakage may overlay or contain an artesian water table. Or such a table may be partially confined (Figure A-5). In both cases, as the body is recharged from surface water, *hydrostatic* pressure rises. If a confining layer is opened by digging or drilling, the water rises above the table 1 foot in altitude for each pressure increase of 62.4 pounds per square foot.

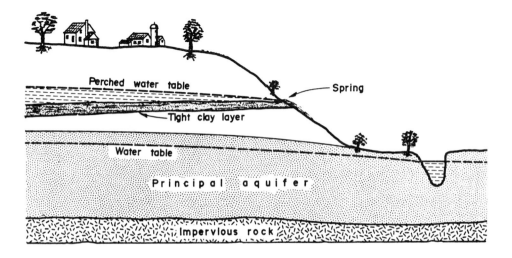

*Figure A-5. Perched Water Table*

*This type of water table occurs above impervious rock or soil and above and separated from the main water table.*

Source: *Ground Water and Wells*, ed. by G. F. Briggs and A. G. Fiedler (St. Paul, Minn.: Wheelabrator Engineered Systems—Johnson Screens, Inc., 1975), p. 21, fig. 12.

In various climates, water tables tend to rise during precipitation periods and fall during dry spells. In humid climates the usual cycle has higher tables during winter and spring and lower tables during summer and into fall. This cycle occurs because cold weather reduces evaporation from precipitation. Colder climates also have reduced transpiration in winter months because so few leaves remain on the deciduous trees, and most plants are dormant.

All three water tables tend to disappear from various locations during dry weather, but you may find normal and artesian tables flowing or seeping at a lower elevation in the form of a *spring* or *seep*. The difference between a spring and a seep is their flow amounts and rates, with a spring being larger and more likely to be constant. The distinction between them is not clear-cut, however.

Springs and seeps can occur at normal and artesian water tables, but a perched table is less likely to be fed from underneath like the other two. Many surface bodies of water, such as lakes and streams, may be fed from both the surface and subsurface.

If a perched table evaporates, only its residues, such as mud and muck or dried soil, may be visible. A perched table that disappears may fool you into thinking it is gone forever. However, it will return when enough precipitation falls and accumulates.

### Aquifers, Aquitards, Aquicludes, and Aquifuges

Of these four geologic formations, aquifers are most important to builders and remodelers. An aquifer is a subsurface course that supplies enough water to be of economic use. This definition is about the only one of the four agreed upon by most authorities. Furthermore, it is the term that most people use to label water flowing underground when they do not know its source and extent. To illustrate economic use, you might say an aquifer has enough flow to supply a residential well with a flow of at least 3 gallons per minute (GPM). But certain jurisdictions require a greater flow for a home. At the other extreme are those aquifers, such as the Oglala in the Western United States, that provide water to large areas.

Aquifers continue to flow so long as they are recharged at the surface. They can flow by gravity underground or flow under or between confining layers, and thus are under pressure greater than that imposed by the atmosphere.

An aquitard is an underground area of soil or rock that contains little water. Some people define an aquitard as having no water, and others define it as being a junior aquifer with such a small amount of flow as to be useless. An aquiclude contains water like an aquifer, but does not yield it easily. An aquifuge is dry.

## *Chemical and Physical Properties of Water*

Even young school children know a water molecule consists of one atom of oxygen and two atoms of hydrogen. Because of this chemical compounding, water molecules have an affinity for each other known as hydrogen bonding. Bonding is greatest in the solid state and least in the gaseous state.

Water weighs 62.4 pounds per cubic foot at 50 degrees Fahrenheit (10 degrees Celsius.) This weight is the standard for judging the weights of most other substances.

The density of water varies with its temperature and contents. Density decreases as temperature rises so hot water flows easier than cold. Because saltwater is denser than fresh, you float higher in the ocean than you do in your local lake. For practical purposes water is incompressible (compared to air for example) so pressure usually does not figure into density calculations. Because water is 821 times heavier than air, the relationship between water and air helps to explain why—

- Water moves underground when the soil is unsaturated.

- Outside a foundation, water "looks" for a hole or crack to gain entry inside the home.
- Rain blows into a home when the windows are left open during a thunder storm.

*Pressure* is the intensity that a specific weight or force exerts over an area. A volume of water 1 foot high and 1 foot square (a cubic foot) weighs 62.4 pounds. If this cubic foot of water is in a container that is open at the top, water imposes pressures against the sides of the container in a more neatly calculated manner than does soil (see Soils). If the water in the container is 10 feet high, the pressure at the bottom, excluding atmospheric pressure, is 624 pounds per square foot. At the top the pressure is zero. The pressures at the sides are functions of the height of the column of water at specific locations. For example, at 1 foot from the bottom, the pressure down and to the sides is 9 multiplied by 62.4 or 561.6 pounds per square foot: 9 X 62.4 = 561.6 At 5 feet from the bottom, the pressure down and to the sides is 5 times 62.4 or 312 pounds per square feet: 5 X 62.4 = 312. The pressures keep reducing as the location of the calculation rises until the pressure becomes zero at the top in all directions.

Another measure of pressure called head directly expresses the altitude, height, or elevation above a base in feet or the equivalent. You can use head to compare differences among heights of water columns.

Vapor pressure is a measure of the gas pressure existing in a wet or damp environment, such as inside a wall, under a slab, or inside a room. This pressure is a function of the relative humidity that in turn depends on temperature. Along with water flow and capillarity, vapor pressure is another means by which moisture moves in and out of homes.

Capillarity requires surface tension and adhesion as was noted in Chapter 2. During capillarity the particles are under less than atmospheric pressure because of the tensile forces in action. In another situation, surface tension combines with gravity to explain why water sheeting down a home's skin penetrates through holes. Without any wind the gravity causes the downward flow. Researchers have estimated surface tension to be in the many thousands of pounds per square inch. It is strong enough to "pull" the water back from the face of a protrusion, such as a sill without a drip edge, toward the wall. If the wall has an opening right there, the water flows in. When a wind blows against a building, a rule of thumb says that water flows up a wall 1 inch for every 10 miles per hour of wind velocity at a specific height.

# Glossary

**absorption**—The inclusion of water into a compound or material.

**adhesion**—The permanent or temporary attachment of one material to another.

**adsorption**—The surface attachment of water to another material.

**aggregate**—Natural sand, gravel, and crushed stone.

**air handler**—The machine containing the fan or blower of a forced-air system that heats and/or cools finished spaces.

**allowable bearing capacity**—The maximum pressure from a foundation to which a given soil or rock should be subjected.

**ambience**—The conditions existing outside a particular space or beyond an object. For example, ambience may refer to temperature, pressure, light, wind, and the like.

**aquiclude**—An underground formation that contains inaccessible water or that transmits it too slowly for use as a water source..

**aquifer**—An underground water source that provides enough water to serve wells.

**aquifuge**—An underground area of soil or rock that contains no water.

**aquitard**—An underground area of soil or rock that contains little water.

**artesian system**—See *watertable, artesian*.

**at rest**—The pressure of soil or soil containing water against a structure, such as a retaining wall, when soil is not moving toward nor away from the structure.

**bars**—The deposits in a stream or river that build up as flowing water drops the load it carries in suspension.

**bed load**—The materials pushed, rolled, and dragged along the bottom of a stream or river.

**benchmark**—A permanent or temporary elevation point set on a fixed object. It is used in surveying as a reference for establishing more elevations and grades.

**bentonite**—A highly expansive clay frequently found in the Rocky Mountains. It is used for waterproofing heavy construction and for drilling mud to reduce friction on the drill, but houses should not be built on it without taking the special precautions as cited above under clay.

**berm**—An embankment, usually of soil (see *dike*).

**binder (soil)**—The element of a soil that exerts cohesion among the particles.

**blind drain**—A subsurface drainage system with its inlet below the surface of the ground and covered by sod, gravel, or a permeable fabric.

**bull's liver**—A "quicksilt" that can flow when saturated and vibrated (see *quick*).

**capillary tension (capillarity)**—One of the scientific concepts that describe the attraction that water molecules have for each other. The attractive force between water molecules ascribed to hydrogen bonding.

**cap or coping**—The metal cover that protects a parapet or other wall from precipitation.

**capping (hydraulic)**—The process of covering a spring head with a collection and conveyance system that transmits the issuing water to a safe outlet.

**check dam**—The low barricade installed in a ditch to slow the water flow by obstructing it.

clay—(1) The finest soil particle, microscopic in size with a diameter smaller than 0.075 mm. or 3 mils (0.003 inch). (2) The family of soils that is sticky and moldable when wet and extremely hard when dry.

clogging—The restriction to the passage of water through an aggregate filter or a geosynthetic material (filter fabric) that results from the migration of soil or rock fines in water flow or seepage.

coating—The covering of a material by a liquid to create a dampproofing or waterproofing membrane on the surface (see *sealant*).

cohesion—The physical and/or chemical attraction soil particles have for each other, usually in clays. The basis for clay's strength.

cold joint—The unintentional separation or gap in a concrete wall between areas that were placed at different times.

collar joint—The vertical space between the back of a brick facing and the adjacent back-up masonry.

confining layer—The underground layer of soil or rock above and/or below a watercourse that prevents water from percolating up or down.

connate water— The water trapped in rocks by geologic action.

consistency (soil)—A soil's resistance to a change in shape.

construction joint—The intentional separation between two sections of a wall or a slab that were placed at different times.

counter flashing—The metal cover that partially overlies the base flashing over the separation between two dissimilar materials. It is fastened to one of the materials and overlaps but is not attached to the other half of the flashing combination (see *step flashing*),

cricket—The small roof, usually A-shaped, behind a barrier to water flow down a roof slope. A typical barrier is a chimney with its long dimension perpendicular to the slope.

dampproofing—A water-repellent process usually applied to the outside of a building, such as a foundation wall, to protect against flowing and shallow-standing water but not deep-standing water. Examples are parging and bituminous coating (see *waterproofing*).

daylighting—The outlet of an underground pipe at the surface from where its flow will continue downslope.

decomposition (geologic)—The chemical weathering of rocks.

deposition—The process whereby materials being carried by water, wind, ice, and the like are dropped onto the surface they are flowing over.

detention—The temporary holding-back or containment of storm water by a pond, basin, or similar device (see *retention*).

dewatering—The general name for removing unwanted water from a construction site, a job in progress, or a completed project.

dew point—The temperature in a body of air when any moisture it contains begins to condense as a result of cooling.

differential leveling—The comparison of elevations or grades using a surveying instrument such as a level or transit. For short distances a plumb-rule together with a straight-edge may be used.

dike—A man-made ridge of soil, rock, or concrete to direct or contain water. Examples are levees, flood walls, and the upright portion of a diversion.

disintegration (geologic)—The physical or mechanical weathering of rocks.

diversion—The combination of a dike and channel in the shape of a leaning S to intercept and redirect water flowing down a slope. The soil dug out of the channel can be used to build the dike.

diverter—(1) The metal L-shaped strip attached diagonally down a roof's slope to direct runoff from an exposed area beneath the roof's edge to a nearby gutter. (2) A flat piece of metal vertically attached to the outer upper edge of a gutter to prevent spillover of large flows such as from a valley. This device redirects the roof runoff into the gutter.

**drainage board**—A geocomposite stood against a wall or laid on a subgrade to intercept and convey flowing water to a safe outlet.

**drawdown**—The lowering of a water table in a well, under a slab, or in a phreatic zone, usually by pumping.

**drip edge**—The bent-out bottom edge of a flashing to prevent downflowing water from curling around and backward (curl-back) against a building because of surface tension.

**drop-off (set-back)**—The reduction in width of a chimney above the highest fireplace to save labor and materials. Where the wider section meets the narrower one, the masonry forms a rake.

**drumlin**—A hillock of gravel deposited by a glacier and left behind when the glacier retreated.

**drywell**—An underground collector of runoff (roof and/or surface) for temporary storage. It is supposed to slowly release the water into neighboring soil for safe dissipation. In practice it often fills quickly and backs-up into its sources where various damages can occur.

**elevation**—The altitude above mean sea level, or some other datum, of a structural component such as a floor or building corner (see *grade*).

**envelope**—The material, such as gravel, a geosynthetic, or both, that surrounds an underground perforated pipe. It serves three purposes. It works as (a) filter for the fines carried in water seeping or flowing toward the drain, (b) a primary conduit for water before it reaches the pipe, and (c) a secondary conduit to convey water that exceeds the capacity of the pipe.

**erosion**—The process whereby soil and rock are worn away by wind, rain, ice, and temperature changes.

**evapotranspiration**—The release of moisture by trees and plants to the atmosphere. For deciduous species (those that lose their leaves in winter), this action occurs mostly during late spring, summer, and early fall.

**expansion joint**—A separation between two similar materials, such as concrete panels in a slab, to allow for movement ("come and go") in response to temperature changes. To be more descriptive the term would include the word *contraction*.

**expansive (soil)**—A family of soils in various parts of the country that contain a large portion of highly plastic (moisture sensitive) clay. Because they are sensitive to water their volumes increase (swell) or decrease (shrink) as clay is added or subtracted naturally or in the laboratory. The term also connotes *contractive*.

**filter fabric**—One of a family of plastic materials, called geotextiles, that is intended to intercept and hold the soil particles carried in suspension in water flowing or seeping toward a subsurface drainage pipe (see *clogging*). (Currently fines tend to accumulate on and in these fabrics, and they block the flow of water into the pipe. Research to solve this problem is in progress.)

**fines**—The microscopic portion of clay and silt soils with particles less than 0.075 mm (0.003) in diameter.

**flashing**—The general name for materials used to separate other materials. They occur at a joint where a smooth connection would be impossible or impractical. Flashings are made of metals and flexible sheetings (see *counter flashing* and *step flashing*).

**fragipan**—A layer of soil relatively impervious to the passage of water that was caused by pressure from overlying material or overburden (see *hardpan*).

**French drain**—A surface or subsurface drainage trench or ditch that contains gravel but no pipe. A filter fabric may surround the gravel as an envelope.

**fuller's earth**—A highly plastic clay that is good for absorbing oils and dyes.

**geocomposite**—An in-plane drainage material with a man-made core of waffle-shaped plastic sheets, interconnected foam balls, or plastic wires in the form of a mesh. It may have a filter fabric on the water side of the core face and either another filter fabric or an impermeable sheet on the other side of the core face.

**geogrid**—Open-grid plastic net used for reinforcing soils or for separating different layers in a subgrade, such as under paving.

**geomembrane**—A virtually impermeable sheet of plastic or artificial rubber used for waterproofing, such as against subsurface walls and slabs.

**geosynthetic**—A broad family of man-made plastics made from synthetic polymers such as polyester, polyethylene, polyproplene, and nylon. These materials serve several purposes, such as—

- a filter in drainage
- a waterproofing membrane
- a reinforcement membrane over a paving subgrade or as fibers in concrete
- a protective membrane between different materials
- an erosion and sediment control using fences, mats, and blankets

The term geosynthetic includes geocomposite, geogrid, geomembrane, and geotextile.

**geotextile**—A woven or nonwoven, porous plastic cloth used for drainage, filtration, reinforcement, or separation. It includes the materials used for filter fabrics and silt fences.

**gley**—A soil whose dark color results from an oxygen shortage resulting from frequent water saturations, rather than from an abundance of organic material such as humus.

**grade, graded, grading**—(1) The altitude of the ground at a specific point in the topography. (2) The slope of the ground. (3) The practice of landforming. Some prefixes for these words are *fine-*, *rough-*, and *finish-*. (4) The proportions of different sized soils in a sample. Some descriptive words that precede *graded* are *uniform-*, *gap-*, *well-*, and *poorly*.

**groundwater**—The water under the surface of the ground (subsurface). The name may imply an underground origin, but most subsurface water originated on the surface somewhere sometime.

**gully**—In erosion and sediment control the name for the deepest and widest ditch or channel washed-out by fast-flowing water.

**gumbo**—A sticky mud, often of plastic clay, that is found in the Southwest, and along the Missouri and Mississippi River valleys. Using it for backfill is risky.

**hardpan**—A layer of soil or rock relatively impervious to the passage of water that was caused by the water-induced "cementing" of grains together.

**humus**—The residue of decaying and decayed organic matter that lies on the surface of the ground, especially under trees.

**hydraulic conductivity**—A technical term from hydrology for the permeability of soil, gravel, and rock.

**hydraulic jump**—As water flows through a channel and over a curb its velocity is slowed as it "piles up" behind the curb before it overflows it. This loss of velocity substantially reduces the force of the flowing water, and thus its strength to erode the downstream channel.

**hydraulics**—A branch of engineering that applies various sciences to liquids at rest and in motion. The equipment and devices for moving water are a major application of this knowledge.

**hydrology**—The science that studies water above, on, and below the surface of the earth.

**hydrostatic**—The pressure exerted by a standing body of fluid that increases with depth.

**ice dam**—The obstruction that occurs in and above gutters when snow and ice have melted and refrozen.

**impermeable**—Theoretically an impermeable surface does not allow the passage of fluids, but in practice it may allow such passage.

**impervious**—Has the same meaning as impermeable.

**infiltration**—The penetration of water into the surface of soil, rock, or pavement.

**in-plane**—This refers to the orientation of an object whose major plane parallels the view of an observer. For example, when you are standing on the ground and looking up at a roof with the eave, faces of the roofing, and the ridge in a line with your eye, you are seeing it in-plane.

**interceptor**—A subsurface drain installed at an angle to the flow or seepage of water to collect and convey it to a safe outlet. Sometimes the word means a surface trench (with or without a berm) angled to divert overland flow.

**interstices**—The open spaces (voids) between and among the solids in soil, rock, or pavement.

**laminar flow**—The flow of water in a pipe that is streamlined not turbulent. As a result the pipe walls create less resistance to the water's passage.

**landforming**—Changing the topography of the land surface using earth movers.

**leader**—A synonym for downspout.

**level spreader**—In erosion and sediment control the widening of a channel at the foot of a slope into a level or nearly level outlet that slows the runoff.

**liquefaction**—The loss of strength of a wet soil caused by shaking. It often occurs during an earthquake. The word *quick* may be used before the words *sand* or *silt* to name a susceptible material.

**liquefy**—In its widest sense, turning a solid, a soil, or a gas into a fluid by adding enough liquid. In a narrow sense, it is the precursor of liquefaction. During erosion, when soils with a large silt content are subjected to water flowing through them, they separate and lose their particles to the water. The water carries these particles away in suspension.

**loam**—The name for soil textures containing various combinations of clay, sand, and silt.

**loess**—A soil containing mostly silt deposited in the upper Midwest by glacial winds. One example, is the loess bluffs along the Missouri River between Sioux City, Iowa, and Omaha, Nebraska. This soil demonstrates a peculiar characteristic that is different from most soils in the world. After slumping during and after being rained on, loess often will re-erect itself to an almost vertical face (slope) when the moisture evaporates. Under and around houses you need to keep loess from ponding water that could soften it and lead to settlement.

**mat (raft)**—A concrete slab or pad, usually reinforced, upon which a building is constructed. Builders use mats where underlying soils are too soft for regular footings or too deep for piles, shafts, or poles to reach. A mat or raft distributes the house loads over a far wider area than regular footings so the house can take advantage of an extremely low allowable bearing capacity of only a few hundred pounds per square foot.

**membrane**—A sheeting intended to be dampproof or waterproof, but it is not always so in practice.

**molecular attraction**—The physical and chemical affinity of particles for each other (see *cohesion*).

**muck**—Another name for mud, both of which usually include silt and organics.

**mudslab**—(1) Another name for a mat or raft. (2) A concrete slab from which workers and equipment work during mucky or soft soil conditions.

**negative side**—The side of a wall or floor opposite the side where the water is standing or flowing. For a house the negative side usually is inside. Dampproofing or waterproofing this side of a wall because the opposite side is inaccessible runs the risk of trapping water inside the wall (see *positive side*).

**optimum moisture content (OMC)**—The maximum water content of a soil when it will most readily and fully compact.

**outlet protection**—At the outlet end of a subsurface pipe where it daylights onto soil, you may need to lay a loose rock apron to prevent erosion by the outrushing water. Another outlet protection device consists of a three-sided box of gravel at the end of the outlet to create a hydraulic jump.

**overburden**—(1) The newer soil or rock that was deposited over a preexisting soil or rock thousands or millions of years later. When the newer material is excavated, some soils lose significant strength as they are unloaded. (2) Also the temporary piling of material over a weak and wet underlying soil to compact it.

**parging (pargeting)**— A cementitious coating or grouting over concrete masonry units (CMUs) as a *dampproofing*. Metallic waterproofing is another name for it, but the word *waterproofing* is a technical misnomer.

**peat**—Partially decomposed and putrefied organic matter found in bogs in Canada, Alaska, and elsewhere. Methane gas could be hazardous. Even dewatered, its bearing capacity would be less than minimal.

**percolation**—The movement of water underground in various directions.

**permafrost**—A condition that occurs in extremely cold climates in which the ground stays frozen all year for a number of years. In the winter permafrost extends from the surface down many feet. (The distance depends on climatic and soil variables.) In the summer the ground thaws from the surface down variable distances.

**permeability**—The ability of a material to conduct water.

**pervious**—The opposite of *impermeable*.

**phreatic zone**—The zone of underground water where all voids are filled. Also called the zone of saturation.

**phreatophyte**—A plant or tree that thrives in an overly wet environment.

**piping**—The internal erosion that takes place in a backfill or in an underfill when flowing water boils through a weak soil. One example is the loss of material from behind a retaining wall through the wall.

**pitch**—(1) The slope of a roof. (2) The coal-tar-based bitumen (in contrast to asphalt) formerly much used in built-up roofing and paving.

**pitch pockets**—Metal sleeves through which cables, pipes, and wires penetrated flat roof on old houses. Builders filled the sleeves to the top with pitch. But over the years the pitch slumped, settled, or shrunk and allowed standing water to enter the attic. More modern flashings have supplanted pitch pockets.

**plasticity**—The capacity of a material to be remolded or deformed by external pressure without breaking. At least a minimum amount of internal moisture must be present in soils for them to have plasticity.

**pore pressure**—The pressure exerted by the fluid in the void spaces of a soil or rock against the solid particles.

**porosity**—The ratio of the volume of the voids to the total volume of a particular material, usually expressed as a percentage. It cannot exceed 100 percent (see *void ratio*).

**positive side**—The water side of a wall or floor and the preferred side in most cases for dampproofing or waterproofing (see *negative side*).

**pressure, dynamic**—The force exerted by water flowing against an area expressed as pounds per square inch (psi).

**pressure, static**—The force exerted by water standing over an area usually expressed as pounds per square foot (psf) and calculated by multiplying the water height in feet by 62.4.

**protection board**—The front, pervious panel of a geocomposite that faces a backfill or a rigid material. Its purpose is to transmit water to—but prevent crushing of—the plastic drainageway behind or under it.

**quick**—Before words such as *clay*, *silt*, and *sand*, this term names a condition, not a soil. It refers to the effect of water from any source on the stability of a material. Seepage or hydrostatic pressure causes a susceptible soil to become a liquid. Quick soils are readily erodible. Other symptoms of quick soils include heaves and blisters in the soil as well as boils (rising air bubbles) also called sand boils.

**rain screen**—The principle of allowing blowing precipitation through holes in the cladding of a building but stopping it at an impermeable inner skin from where it flows by gravity to established outlet devices.

**reaction (soil)**—The acid versus base (alkaline) content expressed as pH.

**recharge**—The infiltration of water into the ground from a source at or above the surface. The source may be precipitation or irrigation.

**relief**—(1) The general relationships among topographic features, the "lay of the land." (2) In a drainage system a horizonal or vertical pipe to allow overflow of full pipes below the surface of the ground. The piping system is called a chimney drain when the water flows vertically up to the surface of the ground because the water flow parallels air flow up a chimney. (3) On water heaters and boilers a special valve that opens automatically when the pressure and temperature of the water exceed specified numbers.

**retention**—In storm water management the permanent containment of overland, roof, yard, and pavement runoffs.

**revetment**—A rough-paved surfacing (also called armoring) of a soil slope to prevent, or at least reduce, erosion by runoff. The slope is shallow enough not to require a retaining wall. Examples of materials used include riprap, broken concrete, and boulders. Sometimes a heavy geotextile is laid over the surface before the armoring, but the erosive power of water is such that over a period of time enough sloughing, scouring, and undermining (caving) take place that even massive boulders will move, tear the protective membrane, and allow the water coursing between the armor to attack the slope.

**rill**—A small channel that is cut into a slope by erosive runoff that started as sheet flow.

**riprap**—The large slabs of rock that are laid on a slope to protect it against erosion. They may weigh 150 pounds or more.

**roiling**—The turbulent action of water flowing over a rough surface.

**"rotten" rock**—A construction term for a rock formation that is easily excavated by the teeth of a loader or backhoe.

**sand**—A granular material that results from disintegration and decomposition of rock.

**scour**—The washing away of soil or other backfill from behind or under structures such as catch basins and outlet pipes. If a wall is undermined by flowing water the action is also called caving.

**scupper**—The pipe that drains water from one roof through a wall onto another roof or into a gutter.

**sealant**—The modern term for caulk (see *coating*).

**sedimentation**—The deposition of soil or rock particles onto a surface by slow-moving water.

**sensitive (soil)**—Used in this book as another name for expansive soils that are especially reactive to moisture changes.

**shear strength**—The strength of soils determined by laboratory tests of their particles' resistance to being separated by forces oriented in different directions.

**shoe**—The "elbow" at the end of the downspout that directs water away from the foundation.

**silt**—(1) A soil with microscopic particles that have a diameter less than 0.075 mm (0.0003 in). (2) A soil that crumbles when dry and may flow when wet.

**skin**—The overall covering of a building excluding the roof.

**slab (framed)**—A concrete slab elevated above the ground.

**slab-on-grade (SOG)**—A concrete pad placed directly on the ground.

**sloughing**—An erosive action similar to scour except that it occurs in unpaved channels away from structures.

**spring (seep)**—Water flowing from the ground surface in any direction.

**step (base) flashing**—The metal or plastic membrane that spans the gap between two pieces of material, such as roofing and siding. Usually the membrane is only fastened along one edge, and it requires an overlapping member to accomplish its watertight goal (see *counter flashing*).

**stilling basin**—A water detainer having either walls and a smooth bottom or only a rough bottom, for slowing runoff and thereby reducing erosion downstream (see *detention*).

**structure (soil)**—A soil science classification category of various shapes and combinations of shapes.

**subgrade**—The soil or rock below a paving system or a future structure.

**subsidence**—The settlement of a soil mass resulting from (a) an internal cause such as natural consolidation or (b) an external cause such as the pumping of fluids from the ground. (Water and oil are examples of such fluids.) Sinkholes and "lost ground" are variations of subsidence. Sinkholes occur over abandoned mines, limestone caverns, broken water pipes, leaking underdrains, and the like. One example of lost ground is the spontaneous migration of soil into an excavation.

**subsurface**—Underground.

**surcharge**—The load imposed on a material by adding soil, water, paving, or a structure above it or against it.

**surface tension**—An example of surface tension in action is water standing slightly above the rim of a vessel without overflowing (see *capillary tension*).

**swale**—A channel whose width is several times its depth.

**terrace**—A step or steps cut into a slope to retard and spread out the water flowing down the slope. The steps should slope down and backward not forward.

**texture (textural)**—A classification of soils and rocks according to their particles sizes.

**toe**—The foot of a slope or a wall.

**toewall**—A short wall built at the bottom of a long slope that does not require either a retaining wall (retainer) or a revetment. Such a wall can prevent erosive effects from littering a sidewalk.

**transpiration**—The exhalation of excess water by a plant or tree through its leaves.

**trench drain**—A subsurface drainage system that includes at least pipe.

**turbid (water)**—Cloudy water.

**underdrain**—Another name for an underground drainage system.

**vadose zone**—In groundwater hydrology the zone above the water table that contains both air and water. Also called the zone of aeration.

**vapor pressure**—The pressure exerted by the gas in a compound.

**velocity check**—A temporary check dam to slow the speed of the water flowing in an unpaved ditch.

**viscosity**—The thickness of a liquid or a fluid, its resistance to flow.

**void ratio**—The ratio of the volume of voids to the volume of solids in a given material, usually expressed as a decimal (see porosity).

**wall jack**—An exit cap through an exterior wall that exhausts dryer, range hood, bath fan, and other vents.

**wash**—The covering over the top of a masonry chimney to prevent the infiltration of rain and snow. It should be made of cement not mortar.

**waterproofing**—The system for stopping the flow of water through a building material even when the water is under a deep static pressure or a large dynamic pressure.

**waterstop**—A device for insertion in the construction joint between the panels of a concrete foundation. The historical molded rubber types do not work as well as the modern plastic devices.

**water table, artesian (confined)**—The height of water under a pressure greater than atmospheric. In an artesian well the water level stands above the surface of the body it taps.

**water table, main (apparent, local, or normal)**—The surface of the free water under atmospheric pressure) that stands without rising or falling in an uncased bore hole after a time of adjustment.

**water table, perched**—A surface of free water that stands above the normal watertable because of an impermeable layer between them.

**water table (siding)**—A molded, sloping board that separates the bottom of siding from, for example, the top of the foundation. Its purpose is to direct water flowing down the face of the siding out over the foundation wall. The groove in its underside serves as a drip edge. It may be used in other locations too. While it protects against water penetration into a building, it also tends to decay and split because of standing in water that may pond underneath it. To eliminate the water table some contractors add "throw" to their wood siding installations by doubling the first course of siding, like a starter course of roof shingles.

**wet feet**—The author's term for a home with water still penetrating at three or more corners of the basement floor after roof and yard runoff corrections have been completed.

# Selected Bibliography

Ambrose, James, and Dimitry Vergun. *Simplified Building Design for Wind and Earthquake Forces.* New York: John Wiley and Sons, 1980. 317 pp.

*Architectural Sheet Metal Manual.* 5th ed. Chantilly, Va.: Sheet Metal and Air Conditioning Contractors National Association, 1993. 380 pp.

Baldwin, Helene L., and C. L. McGuinness. *A Primer on Ground Water.* Washington, D.C.: U.S. Geological Survey, 1963. 26 pp.

Brady, Nyle C. *The Nature and Properties of Soils,* 10th ed. New York: Macmillan, 1989. 880 pp.

*Building Deck Waterproofing.* STP 1084. Laura E. Gish, ed. Philadelphia: American Society for Testing and Materials, 1990. 146 pp.

Campbell, Russell H. *Soil Slips, Debris Flows, and Rainstorms in the Santa Monica Mountains and Vicinity, Southern California.* Professional Paper 851. Washington, D.C.: U.S. Geological Survey, 1975. 51 pp.

Cedergren, Harry R. *Seepage, Drainage, and Flow Nets,* 3rd ed. New York: John Wiley and Sons, 1989. 465 pp.

*Design and Construction of Urban Stormwater Management Systems.* ASCE Manuals and Reports of Engineering Practice No. 77 and WEF Manual of Practice FD-20. New York: American Society of Civil Engineers; and Alexandria, Va.: Water Environment Federation, 1992. 724 pp.

Dictionary of Geological Terms, 3rd Ed. New York: Anchor Press and Doubleday, 1976. 576 pp.

*Drainage of Agricultural Land.* National Engineering Handbook Section 16. Washington, D.C.: Soil Conservation Service, U.S. Department of Agriculture, 1971. 447 pp. Available as PB 85176204 from the National Technical Information Service (NTIS), Springfield, Va.

*Drainage Manual.* Rev. Washington, D.C.: Bureau of Reclamation, U.S. Department of the Interior, 1993. 321 pp.

*Engineering Geology Field Manual.* Washington, D.C.: Bureau of Reclamation, U.S. Department of the Interior, 1989. 599 pp.

*Evaluation, Maintenance, and Upgrading of Wood Structures.* New York: American Society of Civil Engineers, 1982. 428 pp.

Fetter, C. W., Jr. *Applied Hydrogeology,* Columbus, Ohio: Charles E. Merrill, 1980. 488 pp.

*Foundations and Earth Structures.* NAVFAC Design Manual 7.2. Alexandria, Va.: Facilities Engineering Command, U.S. Navy, 1982. 253 pp.

Freeze, R. A., and J. A. Cherry. *Groundwater,* Englewood Cliffs, N.J.: Prentice-Hall, 1979. 604 pp.

*Glossary of Soil Science Terms.* Madison, Wisc.: Soil Science Society of America, 1987. 44 pp.

Gordon, James Edward. *The New Science of Strong Materials.* Princeton, N.J.: Princeton Univ. Press, 1984, 288 pp.

Gray, Donald H., and Andrew T. Leiser. *Biotechnical Slope Protection and Erosion Control.* New York: Van Nostrand Reinhold, 1982. 271 pp.

Greenfield, Steven J., and C. K. Shen. *Foundations in Problem Soils,* Englewood Cliffs, N.J.: Prentice Hall, 1992. 240 pp.

Heath, Ralph C., *Basic Ground-Water Hydrology*, Water-Supply Paper 2220. Washington, D.C.: U.S. Geological Survey, 1983. 84 pp.

Johnson, L. K. *Homeowner's Guide to Overcoming Problems with Marine Clay in Fairfax County*. Fairfax, Va.: Soil Science Office, Fairfax County, 1989. 18 pp.

Kubal, Michael T. *Waterproofing the Building Envelope*. New York: McGraw-Hill, 1993. 277 pp.

Labs, Kenneth, et al. *Building Foundation Design Handbook*. Oak Ridge National Laboratory Sub. 86-72143/1. Minneapolis, Minn.: Underground Space Center, Univ. of Minnesota, 1988. 349 pp.

Langbein, W. B., and Kathleen T. Iseri. *General Introduction and Hydrologic Definitions*. Water-Supply Paper 1541-A. Washington, D.C.: U.S. Geological Survey, 1960. 29 pp.

Lohman, S. W., et al. *Definitions of Selected Ground-Water Terms—Revisions and Conceptual Refinements*. Water-Supply Paper 1988. Washington, D.C.: U.S. Geological Survey, 1972. 21 pp.

Lstiburek, Joseph, and John Carmody. *Moisture Control Handbook*. ORNL Sub. 89-SD350/1. Washington, D.C.: Oak Ridge National Laboratory, U.S. Department of Energy, 1991. 247 pp.

Luthin, James N. *Drainage Engineering*, Huntington, N.Y.: Robert E. Krieger, 1978. 281 pp.

Maurice, A. E. *The Wet Basement Manual*. Addison, Ill.: Aberdeen Group, 1993. 62 pp.

McFadden, Terry T., and F. Lawrence Bennett. *Construction in Cold Regions*. New York: John Wiley and Sons, 1991. 640 pp.

Meinzer, Oscar E. *Outline of Ground-Water Hydrology with Definitions*, Water-Supply Paper 494. Washington, D.C.: U.S. Geological Survey, 1923. 71 pp.

_____. *Plants as Indicators of Ground Water*. Water-Supply Paper 577. Washington, D.C.: U.S. Geological Survey, 1927. 95 pp.

*Nature to Be Commanded—Earth-Science Maps Applied to Land and Water Management*. G. D. Robinson and Andrew M. Spieker, eds. Professional Paper 950. Washington, D.C.: U.S. Geological Survey, 1978. 97 pp.

*NRCA Roofing and Waterproofing Manual*. 3rd ed. Rosemont, Ill.: National Roofing Contractors Association, 1990. 724 pp.

Oliver, Alan C. *Dampness in Buildings*. New York: Nichols, 1988. 232 pp.

Oxley, T. A., and E. G. Gobert. *Dampness in Buildings: Diagnosis, Treatment, Instruments*. London, Eng.: Butterworths, 1983. 123 pp.

Panek, Julian R., and John Philip Cook. *Construction Sealants and Adhesives*. 3rd ed. New York: John Wiley and Sons, 1992. 400 pp.

Parker, Harry, and James Ambrose. *Simplified Mechanics and Strength of Materials*. 5th ed. Somerset, N.J.: John Wiley and Sons, 1992. 408 pp.

*PCA Soil Primer*. Rev. Skokie, Ill.: Portland Cement Association, 1992. 40 pp.

Peck, Ralph B., Walter E. Hanson, and Thomas H. Thornburn. *Foundation Engineering*. 2nd ed. New York: John Wiley and Sons, 1974. 514 pp.

Powers, J. Patrick. *Construction Dewatering:* 2nd ed. New York: John Wiley and Sons, 1992. 528 pp.

*Residential Asphalt Roofing Manual*. Rockville, Md.: Asphalt Roofing Manufacturers Association, 1993. 66 pp.

Rouse, Hunter. *Elementary Mechanics of Fluids*. New York: Dover, 1978. 376 pp.

Sacks, Alvin M., "Deep Foundations," *Fine Homebuilding*, August/September 1991. (Newtown, Conn.: Taunton Press), pp. 58-61.

————. "Rx for Basement Water Problems," *Family Handyman*, September 1981. (St. Paul, Minn.), pp. 37-40.

_____. "Curing Wet Basements," *Remodeling*, March 1988. (Washington, D.C.: Hanley-Wood), p. 14.

_____. "Drainage Problems," *Remodeling*, April 1987. (Washington, D.C.: Hanley-Wood), p. 13.

_____. "Flashings," *Remodeling*, July 1991. (Washington, D.C.: Hanley-Wood), p. 68.

_____. "Geosynthetics," *Remodeling*, November/December 1987. (Washington, D.C.: Hanley-Wood), p. 14.

_____. "Leaks," *Remodeling*, July 1989. (Washington, D.C.: Hanley-Wood), p. 24.

Shelton, John S. *Geology Illustrated*. San Francisco, Calif.: W. H. Freeman, 1966. 434 pp.

Smith, Baird M. *Moisture Problems in Historic Masonry Walls: Diagnosis and Treatment*. Washington, D.C.: National Park Service, U.S. Department of the Interior, 1986. 48 pp. (Available from Technical Information Center (DSC-PGT), National Park Service, U.S. Department of Interior, Denver, Colorado.)

"Soil-Plant-Water Relationships," Chapter 1 in *Irrigation*, Sect. 15 of *National Engineering Handbook*. 2nd ed. Washington, D.C.: Soil Conservation Service, U.S. Department of Agriculture, 1991. (Available as PB 85182269 at National Technical Information Service (NTIS) Springfield, Va.)

*Soil Survey Manual*. USDA Handbook No. 18. Washington, D.C.: U.S. Department of Agriculture, 1993. 437 pp.

*Soils Manual*. 4th ed. Manual Series 10. Lexington, Ky.: Asphalt Institute, 1986. 238 pp.

Sowers, George F. *Introductory Soil Mechanics and Foundations: Geotechnical Engineering*, 4th ed. New York: Macmillan, 1979. 621 pp.

Spano, Stephen J., and Douglas N. Isokait. *Residential Drainage: Dealing with Basement and Erosion Problems*. Laurel, Md.: Washington Suburban Sanitary Commission, 1979. 24 pp.

*Standard Classification of Soils for Engineering Purposes (Unified Soil Classification) D 2487-92*. Philadelphia: American Society for Testing and Materials, 1992. 9 pp.

*Standard Guidelines for the Design, Installation, Operation, and Maintenance of Urban Subsurface Drainage: ASCE 12-92, ASCE 13-93, ASCE 14-93*. New York: American Society of Civil Engineers, 1994. 54 pp.

*Standard Practices for Description and Identification of Soils (Visual-Manual Procedure), D 2488-90*. Philadelphia: American Society for Testing and Materials, 1990, 10 pp.

*Standard Terminology Relating to Soil, Rock, and Contained Fluids, D653-90*. Philadelphia: American Society for Testing and Materials, 1990. 21 pp.

Todd, David Keith. *Groundwater Hydrology*, 2nd ed. New York: John Wiley and Sons, 1980. 535 pp.

Truitt, Marcus M. *Soil Mechanics Technology*. Englewood Cliffs, N.J.: Prentice-Hall, 1983. 284 pp.

Untermann, Richard K. *Grade Easy, An Introductory Course in the Principles and Practices of Grading and Drainage*. Washington, D.C.: American Society of Landscape Architects, 1973. 119 pp.

*Virginia Erosion and Sediment Control Handbook*. 3rd ed. Richmond, Va.: Division of Soil and Water Conservation, Virginia Department of Conservation and Recreation, 1992. 800 pp.

*Water in Exterior Building Walls: Problems and Solutions*. STP 1107. T. A. Schwartz, ed. Philadelphia: American Society for Testing and Materials, 1992. 240 pp.

*Waterproofing and Dampproofing Concrete*. Addison, Ill.: Aberdeen Group, 1983. 48 pp.

Wilson, Forrest. *Building Materials Evaluation Handbook*. New York: Van Nostrand Reinhold, 1984. 368 pp.

# Notes

*Chapter 2. Prevention of Foundation Leakage Before Construction*

1. For advice on dealing with this situation see Steven J. Greenfield and C. K. Shen. *Foundations in Problem Soils*, (Englewood Cliffs, N.J.: Prentice Hall, 1992), 240 pp.

2. Ralph Lewis, *Land Buying Checklist* (Washington, D.C.: Home Builder Press, National Association of Home Builders, 1990), 60 pp.

3. For additional information see Tor H. Nilsen, et al., *Relative Slope Stability and Land-Use Planning in the San Francisco Bay Region, California,* Professional Paper 944 (Washington, D.C.: U.S. Geological Survey, 1979), p. 30. For a full discussion of symptoms and mechanisms see *Landslides Analysis and Control*, Special Report 176, ed. by Robert L. Schuster and Raymond J. Krizek (Washington, D. C.: Transportation Research Board, National Academy of Sciences, 1978), 234 pp., especially Chapter 2, David J. Varnes, "Slope Movement Types and Processes," pp. 11-33, and Chapter 3, Harold T. Rib and Ta Liang, "Recognition and Identification," pp. 34-80.

4. For those of you who are interested in more theory, see the Appendix, Soils, Hydraulics, and Hydrology.

# Index

# Increase Your Business Knowledge with These Bestsellers

**Accounting and Financial Management, third edition**—*Emma Shinn*—This revised and expanded edition shows you how to guide and evaluate your company's financial performance, design an accounting system, choose an accountant, and prepare, analyze, and use your financial reports. It includes new chapters for remodelers, developers, and multiproject companies; the complete, up-to-date NAHB Chart of Accounts; and a list of NAHB-approved software vendors.

**Contracts and Liability for Builders and Remodelers**—*NAHB Legal Department*—This expanded bestseller helps remodelers and builders avoid risks and protect against liability with well-written contracts. New chapters cover the contract between remodeler and owner, liability for builders and remodelers, design/build contracts used by remodelers and custom builders, and contracts with other team members.

**Estimating for Home Builders**—*Jerry Householder and John Mouton*—This book teaches you how to develop complete, accurate construction cost estimates and provides time-saving shortcuts. It gives detailed descriptions of all cost factors: subcontracts, materials, labor, tools, supplies and equipment, jobsite costs, overhead, and markup. Conversion tables, illustrations, and more.

**Production Checklist for Builders and Remodelers**—This step-by-step checklist helps you to get your projects done on time, within budget, and at high quality. It gives you day-to-day procedures to confirm start dates, monitor progress, predict completion dates, meet production schedules, and produce a high-quality home every time.

**Residential Concrete, second edition**—Everything you need to know about high-quality concreting—including guidelines for ordering ready-mixed concrete, working with admixtures, forming, jointing, and curing concrete. Shows you how to fix the most common concrete problems with detailed, easy-to-follow remedies.

**Scheduling for Builders**—*Jerry Householder*—Schedule your work more profitably with these easy-to-understand, step-by-step procedures. They will help you calculate construction timetables, forecast costs, and modify schedules when unexpected delays arise. Shows you how to streamline your scheduling process with bar charts, arrow diagrams, and precedence networks.

**Understanding House Construction**—*John Kilpatrick*—This book describes how homes are built from groundbreaking to final inspection. It introduces the methods, materials, and terms used in homebuilding and shows each step of the process with photos, diagrams, and explanations. Winner of the Distinguished Technical Communications Award presented by the Society for Technical Communications.

*To order these books or to receive a current catalog of Home Builder Press products call (800) 223-2665 or write to—*

HOME BUILDER BOOKSTORE

Home Builder Bookstore®
National Association of Home Builders
1201 15th Street, NW
Washington, DC 20005-2800